Christian Ropp

Ropp's Easy Calculator

Christian Ropp

Ropp's Easy Calculator

ISBN/EAN: 9783337389390

Printed in Europe, USA, Canada, Australia, Japan

Cover: Foto ©berggeist007 / pixelio.de

More available books at **www.hansebooks.com**

ROPP'S
EASY CALCULATOR

DESIGNED FOR THE USE OF

FARMERS, MECHANICS, BUSINESS MEN AND LABORERS.

CONTAINING MANY CONVENIENT AND VALUABLE

TABLES SHOWING THE VALUE OF WHEAT, CORN, RYE, OATS, BARLEY, CATTLE, HOGS, HAY, COAL, LUMBER, MERCHANDISE; THE SIMPLE AND COMPOUND INTEREST AT 6, 7, 8 AND 10 PER CENT.; MEASUREMENT OF BOARDS, SCANTLINGS, TIMBERS, SAW LOGS, CISTERNS, TANKS, GRANARIES, CORN-CRIBS, WAGON-BEDS; TIME TABLE, WAGES TABLES, ETC.

ALSO ENTIRELY NEW AND PRACTICAL

METHODS OF RAPID CALCULATION.

By Christian Ropp, Jr.

BLOOMINGTON, ILL.

1879.

NOTE.---The tables and methods embodied in this work are supposed to be absolutely accurate and reliable. Any one who may detect a mathematical error in any of the following tables, will be entitled to one hundred copies of this work, by communicating the fact to the author.

CONTENTS.

PREFACE.

Any invention or discovery that tends to ease and accelerate physical or mental labor, adds to the public welfare, and will be appreciated in this age of intelligence, progress and thrift. A work of this kind which saves both time and labor, has long been wanted, especially by the agricultural community, since so many like the author himself, who is a practical farmer, have had limited advantages for obtaining a proper education.

Nearly all the *practical* features found in Arithmetics, Ready-Reckoners, Lightning-Calculators, Interest, Lumber and Wages tables, are embodied in this work, and in addition it contains a great many *original* rules and tables, which are by far the most valuable part of the work.

The tables are unequalled for clearness and simplicity and will enable *any one*—the least conservant with figures—to become his own accountant almost instantaneously. They show at a glance the accurate value of all kinds of Grain, Stock, Hay, Coal, Lumber and Merchandise, from one pound to a car load, and from the lowest to the highest prices that the market is likely to reach. The simple and compound Interest at 6, 7, 8 and 10 per cent. on all sums from $1 to $2000 and from one day to six years. Measurement of Lumber, Saw Logs, Cisterns, Granaries, Corn-Cribs, Wagon Beds, etc. A Time, Wages, and many other useful tables and important information.

The "Contracted Methods of Calculation" which save a vast amount of figures and mental labor, and which are in vain sought for in any other mathematical work, will be admired by all who appreciate rapidity, brevity and simplicity.

The mechanical part of the work will commend itself and with its silicate slate, memorandum and pocket for papers will be found a most convenient and desirable pocket manual, adapted to all classes of men whether in business or out of business.

That this little volume may prove interesting and profitable to all who consult its pages, is the sincere desire of the

AUTHOR.

BLOOMINGTON, ILL., April, 1876.

THE MULTIPLICATION TABLE

Is inserted for the convenience of those who have not thoroughly committed it to memory.

1	2	3	4	5	6	7	8	9	10	11	12
2	4	6	8	10	12	14	16	18	20	22	24
3	6	9	12	15	18	21	24	27	30	33	36
4	8	12	16	20	24	28	32	36	40	44	48
5	10	15	20	25	30	35	40	45	50	55	60
6	12	18	24	30	36	42	48	54	60	66	72
7	14	21	28	35	42	49	56	63	70	77	84
8	16	24	32	40	48	56	64	72	80	88	96
9	18	27	36	45	54	63	72	81	90	99	108
10	20	30	40	50	60	70	80	90	100	110	120
11	22	33	44	55	66	77	88	99	110	121	132
12	24	36	48	60	72	84	96	108	120	132	144

Read carefully ALL the following

EXPLANATIONS

To Grain, Stock, Hay, Coal, Lumber, Interest Tables, Etc.

EXAMPLES.—Find the value of a load of Wheat weighing 3450 (3000+400+50) lbs. at 48 cts. per bu. Turn to page 7.

Look for the price in the left hand column, and for the weight or quantity in the first 3 lines at the top. Lay the silicate slate with its upper edge directly below the line in which the price is found. Look for 3000 at the top, run down the column till opposite 48, where you will find 2400 ($24.00); write it down it being the value of 3000 lbs. at 48 cts. per bu. In like manner the value of 400 lbs. is found to be 3.20, the *right-hand* cipher not falling *below* the 400, being rejected. The value of 50 lbs. is 40 cts; the *only* figures *vertically* below 50. The three numbers added and two places pointed off from the right, gives the answer in dollars and cents.

3000 lbs.	cost	24.00
400 "	"	3.20
50 "	"	.40
	Ans.	$27.60

For the *thousands* or any number found in the *upper* line take all the figures below, opposite the given price; for the *hundreds* or numbers found in the *second* line, reject the *right hand* figure in the corresponding number below; for the *tens* or numbers found in the *third* line, reject the *two* right hand figures below, etc.

The small figures on the right of the second and third columns are to be used *only* when the weight or quantity is 10 or 20,000. For instance, 10,000 lbs. of wheat at 92 cts. per bu. come to $153.33; 1000 lbs. to $15.33; 100 lbs. to $1.53; 10 lbs. to 15 cts; 1 lb. to 1 ct. 5 mills, practically 2* cts.

When a fraction occurs in the price, find the value for the whole number first, then for the fraction—found near the top of column.

Find the cost of 20,630 lbs. of Corn at 48¾ cts. per bu. (Page 10.)

20,000 lbs. at 48 c. cost	171.43,	at ¾ c.	2.68	equals	174.11	at 48¾ cts.
600 " " " "	5.14	" "	8	=	5.22	" " "
30 " " " "	.26	" "	0	=	.26	" " "
				Ans.	$179.59	

The *fourth line* from top (old style figures) at the beginning of each grain table, shows the *weight* reduced to *bushels* and *hundredths.*

How many bushels in the above car load of Corn weighing 20,630 lbs.?

In 20,000 lbs. there are 357 bu. and 14 hundredths; in 600 lbs., 10 bu. and 71 hundredths; in 30 lbs., 54 hundredths. In all 368 bushels and 39 hundredths of a bushel.

20,000 lbs.	equal	357.14 bu.
600 "	=	10.71 "
30 "	=	.54 "
	Ans.	368.39 "

What is the cost of a load of Oats weighing 2410 lbs. at 42 cts. per bu.? (Page 13.)

2000 lbs.	cost	26 25
400 "	"	5.25
10 "	"	.13
	Ans.	$31.63

How much does a load of Hay weighing 1670 lbs. come to at $11.50 per ton? (Page 17.)

1000 lbs.	cost	5.75
600 "	"	3.45
70 "	"	.40
	Ans.	$9.60

Find the value of a lot of Hogs weighing 4060 lbs. at $4.85 cts. per hundred. (Page 19.) Ans. $196.91

First find the value at $4.75 then at 10 cts. per hundred.

*NOTE.—When the nearest rejected figure is 5 or over, add 1 to the part retained; thus, counting *one half* or over, a *whole one*, and rejecting what is *under*.

Interest Tables.—The Time in years, months and days will be found in left hand column; the Principal, from $1 to $2000, at the top of each page.

Find the Interest of $300 for 2 yrs. 5 mos. 14 da. at 6 per ct. (Page 20.)

Look for 300 at the top, run down the column where you will find the interest for 14 days to be 70 cts., for 5 months $7.50, and for 2 years $36.00. The three numbers added give the required interest.

Int. for	14 da.	.70
" "	5 mos.	7.50
" "	2 yrs.	36.00
	Ans.	$44.20

Find the Int. on 94.50 for 7 mos. and 23 da. at 7 per cent.

When "cents" are given, find the interest for as many dollars and take the *hundredth* part thereof.

Int. on $90 for 7 mo. 23 da.	4.08
" " 4 " " "	.18
" " .50 " "	2
Ans.	$4.28

Reject as many figures below as there are ciphers omitted above. For instance, the interest of $1000 for 93 days at 10 per cent. is $25.83; of $100, $2.58; of $10, 26* cts.; of $1, 3* cts., etc.

The 5 lower lines show the Compound Interest. The compound interest of $500 for 6 years at 8 per cent. is $293.44.

The Time Table—(Page 24.) for ascertaining the date on which a note or bill matures when given for a certain number of days. Also, for finding the exact time intervening between two dates.

When will a note drawn on the 25th of Jan. for 100 days become due?

The 25th of Jan. is likewise the 25th day of the year; adding 100 to this makes 125, which will be found opposite May, below 5. With 3 days of grace it would mature May 8.

How many days, and how many weeks, from Oct. 19 to March 1?

Subtract 292 from 425. The 19th of Oct. is the 292nd day of the year, the 1 of March—of the following year—the 425th. The difference (133) is the time in days which divided by 7, gives it in weeks.

425
292
133 da.

In passing over the 29 of February, make allowance for 1 day more.

To find the Time in months, years and days, see rule page 47.

Lumber Table.—(Page 25.) Look for the width of boards, or for the width and thickness of timbers, in the left hand columns, and for the length at the top—in the angle will be found the contents in square feet and inches. Thus, a board 17 inches wide, and 18 feet long contains 25 ft. and 6 in. A sill 8 by 8 and 22 ft. long, 117 ft. 4 in., etc.

Log Table.—(Page 26.) Find the diameter in left hand column, the length at the top. A saw log 19 inches in diameter and 18 ft. long, will make 254 ft. of square-edged inch boards.

Cistern Table.—(Page 26.) The mean or average diameter will be found in left hand column, the depth at the top. A well 4 feet in diameter and 20 ft. deep will hold 60 barrels. (31½ gallons to the barrel). A cistern 7½ ft. in diameter and 12 ft. deep contains 126 barrels. A tank or cistern, in order to have a capacity of 504 barrels must be built 15 ft. in diameter and 12 ft. deep, or 13 ft. in diameter and 16 feet deep.

Wages Table.—(Page 28.) Look for the rate per week or month in top lines, for the hours and days, in left hand column. Thus, at $7.50 per week, a person would earn $5 in 4 days, and 75 cts. in 6 hours. (10 hours a day's work.) At $20 a month, a man earns 77 cts. a day; $8.46 in 11 days, etc. (26 working days a month.

For Explanations to Tables showing contents of Granaries and Corn-Cribs, see page 27.

*See note page 4.

The Table on page 80 embodies nearly all the features found in "Ready Reckoners," and is handy for finding the value of Butter, Eggs, Goods, etc. The price is found in left hand column, the quantity in upper line, or vice versa. Thus, 23 lbs. of butter at 35 cts. are worth $8.05. 28 yds. of goods at 45 cts. come to $12.60, etc.

When the given price or quantity is *less* or *exceeds* the extremes found in the tables, *double*, or *take half* of some convenient number. For instance, 46 lbs. would cost *twice* as much as 23 lbs., and at 15 cts. would amount to only *half* as much as at 30 cts.

"Contracted Method of Multiplication."—(See Pages 35 & 36.)

There are usually from *two* to *five times* as many figures involved in the ordinary methods of calculation, as are required, by involving common or decimal fractions which fall *below cents* or *hundredths*—the lowest order regarded in *practical* calculations. All this labor and useless figuring is avoided by the following simple and *scientific* principle, which is the chief element embodied in the rules for finding the value of grain, stock, merchandise, etc., pages 40 and 42; for computing interest pages 48 and 51; for ascertaining the capacity of granaries, corn-cribs, cisterns, tanks, etc., pages 67 and 68; besides many others.

Find the cost of 94¾ (94.75) yds. of goods at 83¾ (.8375) cts. per yd.

Write the common fractions decimally.

Write one of the terms in *reversed* order under the other, so that its figure on the *right* of the decimal point will fall *below* the *same figure* in the upper term. Or so that *tenths* fall under *tenths, hundreths* under *units*, etc. No order lower than "cents" will then arise in the products.

```
        9 4.7 5
      5 7 3 8
      -------
      7 5 8 0
        2 8 4
          6 6
            5
         ----
Ans. $7 9.3 5
```

In multiplying by the 8, commence with the 7 above it and proceed in the usual manner, adding, however, the (4) *tens* from the *rejected* figure 5, (8 times 5.) In multiplying by the 3, begin with the 4 above it and add the (2) *tens* from the *nearest* rejected figure 7, (3 times 7). Coming to the 7, multiply the 9 above it and add the (3*) *tens* from the *nearest* rejected figure 4, (7 times 4). There being no figure above the 5, simply multiply the *nearest* rejected figure 9 by it (mentally), and set the (5*) *tens* in the right hand column. Write the first figure of each partial product in the same column. Add and point off two places—the result will be correct within a few mills.

*In carrying tens from the product of the nearest rejected figure, carry *one more* when the *units* figure of the product is *five* or *over*. For instance, from 5 to 14 carry *one*; from 15 to 24 carry *two*; from 25 to 34 carry *three*; from 35 to 44 carry *four*, etc. By this principle *one-half* or over, is counted a *whole one*, and what is under is rejected. Thus one eqalizes the other.

TABLE showing the number of Pounds to the Bushel,

As recognized by the Laws of the United States.

Wheat	60	Hung'n Grass Seed	45	Apples, Green	56
Corn, shelled	56	Blue Grass Seed	14	Dried Apples	24
Corn, in the ear	70	Millet Seed	50	Dried Peaches	33
Rye	56	Red Top Seed	14	Cornmeal	48
Oats	32	White Beans	60	Bran	20
Barley	48	Castor Beans	46	Malt	38
Buckwheat	52	Peas	60	Stone Coal	80
Timothy Seed	45	Potatoes	60	Charcoal,	22
Clover Seed	60	Sweet Potatoes	55	Salt	65
Flax Seed	56	Onions	57	Lime, unslacked	80
Hemp Seed	44	Turnips	55	Plastering Hair	8

A Bushel contains 2150.4 cubic inches. A Gallon 231. A Box 13 by 13 inches and 12¾ inches deep contains a bushel, or 2154¾ cu. in.

Weight.	1000^{0} 100 10	2000^{0} 200 20	3000 300 30	4000 400 40	5000 500 50	6000 600 60	7000 700 70	8000 800 80	9000 900 90
Bush.	1666^{7}	3333^{3}	5000	6667	8333	10000	11667	13333	15000
Price per Bu. ¼	4^{2}	8^{3}	13	17	21	25	29	33	38
⅓	5^{6}	11^{1}	17	22	28	33	39	44	50
½	8^{3}	16^{7}	25	33	42	50	58	67	75
⅔	11^{1}	22^{2}	33	44	56	67	78	89	100
¾	12^{5}	25^{0}	38	50	63	75	88	100	113
.48	800^{0}	1600^{0}	2400	3200	4000	4800	5600	6400	7200
.49	816^{7}	1633^{3}	2450	3267	4083	4900	5717	6533	7350
.50	833^{3}	1666^{7}	2500	3333	4167	5000	5833	6667	7500
.51	850^{0}	1700^{0}	2550	3400	4250	5100	5950	6800	7650
.52	866^{7}	1733^{3}	2600	3467	4333	5200	6067	6933	7800
.53	883^{3}	1766^{7}	2650	3533	4417	5300	6183	7067	7950
.54	900^{0}	1800^{0}	2700	3600	4500	5400	6300	7200	8100
.55	916^{7}	1833^{3}	2750	3667	4583	5500	6417	7333	8250
.56	933^{3}	1866^{7}	2800	3733	4667	5600	6533	7467	8400
.57	950^{0}	1900^{0}	2850	3800	4750	5700	6650	7600	8550
.58	966^{7}	1933^{3}	2900	3867	4833	5800	6767	7733	8700
.59	983^{3}	1966^{7}	2950	3933	4917	5900	6883	7867	8850
.60	1000^{0}	2000^{0}	3000	4000	5000	6000	7000	8000	9000
.61	1016^{7}	2033^{3}	3050	4067	5083	6100	7117	8133	9150
.62	1033^{3}	2066^{7}	3100	4133	5167	6200	7233	8267	9300
.63	1050^{0}	2100^{0}	3150	4200	5250	6300	7350	8400	9450
.64	1066^{7}	2133^{3}	3200	4267	5333	6400	7467	8533	9600
.65	1083^{3}	2166^{7}	3250	4333	5417	6500	7583	8667	9750
.66	1100^{0}	2200^{0}	3300	4400	5500	6600	7700	8800	9900
.67	1116^{7}	2233^{3}	3350	4467	5583	6700	7817	8933	10050
.68	1133^{3}	2266^{7}	3400	4533	5667	6800	7933	9067	10200
.69	1150^{0}	2300^{0}	3450	4600	5750	6900	8050	9200	10350
.70	1166^{7}	2333^{3}	3500	4667	5833	7000	8167	9333	10500
.71	1183^{3}	2366^{7}	3550	4733	5917	7100	8283	9467	10650
.72	1200^{0}	2400^{0}	3600	4800	6000	7200	8400	9600	10800
.73	1216^{7}	2433^{3}	3650	4867	6083	7300	8517	9733	10950
.74	1233^{3}	2466^{7}	3700	4933	6167	7400	8633	9867	11100
.75	1250^{0}	2500^{0}	3750	5000	6250	7500	8750	10000	11250
.76	1266^{7}	2533^{3}	3800	5067	6333	7600	8867	10133	11400
.77	1283^{3}	2566^{7}	3850	5133	6417	7700	8983	10267	11550
.78	1300^{0}	2600^{0}	3900	5200	6500	7800	9100	10400	11700
.79	1316^{7}	2633^{3}	3950	5267	6583	7900	9217	10533	11850
.80	1333^{3}	2666^{7}	4000	5333	6667	8000	9333	10667	12000
.81	1350^{0}	2700^{0}	4050	5400	6750	8100	9450	10800	12150
.82	1366^{7}	2733^{3}	4100	5467	6833	8200	9567	10933	12300
.83	1383^{3}	2766^{7}	4150	5533	6917	8300	9683	11067	12450
.84	1400^{0}	2800^{0}	4200	5600	7000	8400	9800	11200	12600
.85	1416^{7}	2833^{3}	4250	5667	7083	8500	9917	11333	12750
.86	1433^{3}	2866^{7}	4300	5733	7167	8600	10033	11467	12900
.87	1450^{0}	2900^{0}	4350	5800	7250	8700	10150	11600	13050
.88	1466^{7}	2933^{3}	4400	5867	7333	8800	10267	11733	13200
.89	1483^{3}	2966^{7}	4450	5933	7417	8900	10383	11867	13350
.90	1500^{0}	3000^{0}	4500	6000	7500	9000	10500	12000	13500
.91	1516^{7}	3033^{3}	4550	6067	7583	9100	10617	12133	13650
.92	1533^{3}	3066^{7}	4600	6133	7667	9200	10733	12267	13800
.93	1550^{0}	3100^{0}	4650	6200	7750	9300	10850	12400	13950
.94	1566^{7}	3133^{3}	4700	6267	7833	9400	10967	12533	14100

Weight	1000 100 10	2000 200 20	3000 300 30	4000 400 40	5000 500 50	6000 600 60	7000 700 70	8000 800 80	9000 900 90
.95	1583^{3}	3166^{7}	4750	6333	7917	9500	11083	12667	14250
.96	1600^{0}	3200^{0}	4800	6400	8000	9600	11200	12800	14400
.97	1616^{7}	3233^{3}	4850	6467	8083	9700	11317	12933	14550
.98	1633^{3}	3266^{7}	4900	6533	8167	9800	11433	13067	14700
.99	1650^{0}	3300^{0}	4950	6600	8250	9900	11550	13200	14850
1.00	1666^{7}	3333^{3}	5000	6667	8333	10000	11667	13333	15000
1.01	1683^{3}	3366^{7}	5050	6733	8417	10100	11783	13467	15150
1.02	1700^{0}	3400^{0}	5100	6800	8500	10200	11900	13600	15300
1.03	1716^{7}	3433^{3}	5150	6867	8583	10300	12017	13733	15450
1.04	1733^{3}	3466^{7}	5200	6933	8667	10400	12133	13867	15600
1.05	1750^{0}	3500^{0}	5250	7000	8750	10500	12250	14000	15750
1.06	1766^{7}	3533^{3}	5300	7067	8833	10600	12367	14133	15900
1.07	1783^{3}	3566^{7}	5350	7133	8917	10700	12483	14267	16050
1.08	1800^{0}	3600^{0}	5400	7200	9000	10800	12600	14400	16200
1.09	1816^{7}	3633^{3}	5450	7267	9083	10900	12717	14533	16350
1.10	1833^{3}	3666^{7}	5500	7333	9167	11000	12833	14667	16500
1.11	1850^{0}	3700^{0}	5550	7400	9250	11100	12950	14800	16650
1.12	1866^{7}	3733^{3}	5600	7467	9333	11200	13067	14933	16800
1.13	1883^{3}	3766^{7}	5650	7533	9417	11300	13183	15067	16950
1.14	1900^{0}	3800^{0}	5700	7600	9500	11400	13300	15200	17100
1.15	1916^{7}	3833^{3}	5750	7667	9583	11500	13417	15333	17250
1.16	1933^{3}	3866^{7}	5800	7733	9667	11600	13533	15467	17400
1.17	1950^{0}	3900^{0}	5850	7800	9750	11700	13650	15600	17550
1.18	1966^{7}	3933^{3}	5900	7867	9833	11800	13767	15733	17700
1.19	1983^{3}	3966^{7}	5950	7933	9917	11900	13883	15867	17850
1.20	2000^{0}	4000^{0}	6000	8000	10000	12000	14000	16000	18000
1.21	2016^{7}	4033^{3}	6050	8067	10083	12100	14117	16133	18150
1.22	2033^{3}	4066^{7}	6100	8133	10167	12200	14233	16267	18300
1.23	2050^{0}	4100^{0}	6150	8200	10250	12300	14350	16400	18450
1.24	2066^{7}	4133^{3}	6200	8267	10333	12400	14467	16533	18600
1.25	2083^{3}	4166^{7}	6250	8333	10417	12500	14583	16667	18750
1.26	2100^{0}	4200^{0}	6300	8400	10500	12600	14700	16800	18900
1.27	2116^{7}	4233^{3}	6350	8467	10583	12700	14817	16933	19050
1.28	2133^{3}	4266^{7}	6400	8533	10667	12800	14933	17067	19200
1.29	2150^{0}	4300^{0}	6450	8600	10750	12900	15050	17200	19350
1.30	2166^{7}	4333^{3}	6500	8667	10833	13000	15167	17333	19500
1.31	2183^{3}	4366^{7}	6550	8733	10917	13100	15283	17467	19650
1.32	2200^{0}	4400^{0}	6600	8800	11000	13200	15400	17600	19800
1.33	2216^{7}	4433^{3}	6650	8867	11083	13300	15517	17733	19950
1.34	2233^{3}	4466^{7}	6700	8933	11167	13400	15633	17867	20100
1.35	2250^{0}	4500^{0}	6750	9000	11250	13500	15750	18000	20250
1.36	2266^{7}	4533^{3}	6800	9067	11333	13600	15867	18133	20400
1.37	2283^{3}	4566^{7}	6850	9133	11417	13700	15983	18267	20550
1.38	2300^{0}	4600^{0}	6900	9200	11500	13800	16100	18400	20700
1.39	2316^{7}	4633^{3}	6950	9267	11583	13900	16217	18533	20850
1.40	2333^{3}	4666^{7}	7000	9333	11667	14000	16333	18667	21000
1.41	2350^{0}	4700^{0}	7050	9400	11750	14100	16450	18800	21150
1.42	2366^{7}	4733^{3}	7100	9467	11833	14200	16567	18933	21300
1.43	2383^{3}	4766^{7}	7150	9533	11917	14300	16683	19067	21450
1.44	2400^{0}	4800^{0}	7200	9600	12000	14400	16800	19200	21600
1.45	2416^{7}	4833^{3}	7250	9667	12083	14500	16917	19333	21750
1.46	2433^{3}	4866^{7}	7300	9733	12167	14600	17033	19467	21900
1.47	2450^{0}	4900^{0}	7350	9800	12250	14700	17150	19600	22050

Weight.	1000 100 10	2000 200 20	3000 300 30	4000 400 40	5000 500 50	6000 600 60	7000 700 70	8000 800 80	9000 900 90
1.48	2466^{7}	4933^{3}	7400	9867	12333	14800	17267	19733	22200
1.49	2483^{3}	4966^{7}	7450	9933	12417	14900	17383	19867	22350
1.50	2500^{0}	5000^{0}	7500	10000	12500	15000	17500	20000	22500
1.51	2516^{7}	5033^{3}	7550	10067	12583	15100	17617	20133	22650
1.52	2533^{3}	5066^{7}	7600	10133	12667	15200	17733	20267	22800
1.53	2550^{0}	5100^{0}	7650	10200	12750	15300	17850	20400	22950
1.54	2566^{7}	5133^{3}	7700	10267	12833	15400	17967	20533	23100
1.55	2583^{3}	5166^{7}	7750	10333	12917	15500	18083	20667	23250
1.56	2600^{0}	5200^{0}	7800	10400	13000	15600	18200	20800	23400
1.57	2616^{7}	5233^{3}	7850	10467	13083	15700	18317	20933	23550
1.58	2633^{3}	5266^{7}	7900	10533	13167	15800	18433	21067	23700
1.59	2650^{0}	5300^{0}	7950	10600	13250	15900	18550	21200	23850
1.60	2666^{7}	5333^{3}	8000	10667	13333	16000	18667	21333	24000
1.61	2683^{3}	5366^{7}	8050	10733	13417	16100	18783	21467	24150
1.62	2700^{0}	5400^{0}	8100	10800	13500	16200	18900	21600	24300
1.63	2716^{7}	5433^{3}	8150	10867	13583	16300	19017	21733	24450
1.64	2733^{3}	5466^{7}	8200	10933	13667	16400	19133	21867	24600
1.65	2750^{0}	5500^{0}	8250	11000	13750	16500	19250	22000	24750
1.66	2766^{7}	5533^{3}	8300	11067	13833	16600	19367	22133	24900
1.67	2783^{3}	5566^{7}	8350	11133	13917	16700	19483	22267	25050
1.68	2800^{0}	5600^{0}	8400	11200	14000	16800	19600	22400	25200
1.69	2816^{7}	5633^{3}	8450	11267	14083	16900	19717	22533	25350
1.70	2833^{3}	5666^{7}	8500	11333	14167	17000	19833	22667	25500
1.71	2850^{0}	5700^{0}	8550	11400	14250	17100	19950	22800	25650
1.72	2866^{7}	5733^{3}	8600	11467	14333	17200	20067	22933	25800
1.73	2883^{3}	5766^{7}	8650	11533	14417	17300	20183	23067	25950
1.74	2900^{0}	5800^{0}	8700	11600	14500	17400	20300	23200	26100
1.75	2916^{7}	5833^{3}	8750	11667	14583	17500	20417	23333	26250
1.76	2933^{3}	5866^{7}	8800	11733	14667	17600	20533	23467	26400
1.77	2950^{0}	5900^{0}	8850	11800	14750	17700	20650	23600	26550
1.78	2966^{7}	5933^{3}	8900	11867	14833	17800	20767	23733	26700
1.79	2983^{3}	5966^{7}	8950	11933	14917	17900	20883	23867	26850
1.80	3000^{0}	6000^{0}	9000	12000	15000	18000	21000	24000	27000
1.81	3016^{7}	6033^{3}	9050	12067	15083	18100	21117	24133	27150
1.82	3033^{3}	6066^{7}	9100	12133	15167	18200	21233	24267	27300
1.83	3050^{0}	6100^{0}	9150	12200	15250	18300	21350	24400	27450
1.84	3066^{7}	6133^{3}	9200	12267	15333	18400	21467	24533	27600
1.85	3083^{3}	6166^{7}	9250	12333	15417	18500	21583	24667	27750
1.86	3100^{0}	6200^{0}	9300	12400	15500	18600	21700	24800	27900
1.87	3116^{7}	6233^{3}	9350	12467	15583	18700	21817	24933	28050
1.88	3133^{3}	6266^{7}	9400	12533	15667	18800	21933	25067	28200
1.89	3150^{0}	6300^{0}	9450	12600	15750	18900	22050	25200	28350
1.90	3166^{7}	6333^{3}	9500	12667	15833	19000	22167	25333	28500
1.91	3183^{3}	6366^{7}	9550	12733	15917	19100	22283	25467	28650
1.92	3200^{0}	6400^{0}	9600	12800	16000	19200	22400	25600	28800
1.93	3216^{7}	6433^{3}	9650	12867	16083	19300	22517	25733	28950
1.94	3233^{3}	6466^{7}	9700	12933	16167	19400	22633	25867	29100
1.95	3250^{0}	6500^{0}	9750	13000	16250	19500	22750	26000	29250
1.96	3266^{7}	6533^{3}	9800	13067	16333	19600	22867	26133	29400
1.97	3283^{3}	6566^{7}	9850	13133	16417	19700	22983	26267	29550
1.98	3300^{0}	6600^{0}	9900	13200	16500	19800	23100	26400	29700
1.99	3316^{7}	6633^{3}	9950	13267	16583	19900	23217	26533	29850
2.00	3333^{3}	6666^{7}	10000	13333	16667	20000	23333	26667	30000

Weight.	1000 100 10	2000 200 20	3000 300 30	4000 400 40	5000 500 50	6000 600 60	7000 700 70	8000 800 80	9000 900 90
Bush.	1785^{7}	3571^{4}	5357	7143	8929	10714	12500	14286	16071
Price per Bu. ¼	4^{5}	8^{9}	13	18	22	27	31	36	40
⅓	6^{0}	11^{9}	18	24	30	36	42	48	54
½	8^{9}	17^{9}	27	36	45	54	63	71	80
⅔	11^{9}	23^{8}	36	48	60	71	83	95	107
¾	13^{4}	26^{8}	40	54	67	80	94	107	121
.20	357^{1}	714^{3}	1071	1429	1786	2143	2500	2857	3214
.21	375^{0}	750^{0}	1125	1500	1875	2250	2625	3000	3375
.22	392^{9}	785^{7}	1179	1571	1964	2357	2750	3143	3536
.23	410^{7}	821^{4}	1232	1643	2054	2464	2875	3286	3696
.24	428^{6}	857^{1}	1286	1714	2143	2571	3000	3429	3857
.25	446^{4}	892^{9}	1339	1786	2232	2679	3125	3571	4018
.26	464^{3}	928^{6}	1393	1857	2321	2786	3250	3714	4179
.27	482^{1}	964^{3}	1446	1929	2411	2893	3375	3857	4339
.28	500^{0}	1000^{0}	1500	2000	2500	3000	3500	4000	4500
.29	517^{9}	1035^{7}	1554	2071	2589	3107	3625	4143	4661
.30	535^{7}	1071^{4}	1607	2143	2679	3214	3750	4286	4821
.31	553^{6}	1107^{1}	1661	2214	2768	3321	3875	4429	4982
.32	571^{4}	1142^{9}	1714	2286	2857	3429	4000	4571	5143
.33	589^{3}	1178^{6}	1768	2357	2946	3536	4125	4714	5304
.34	607^{1}	1214^{3}	1821	2429	3036	3643	4250	4857	5464
.35	625^{0}	1250^{0}	1875	2500	3125	3750	4375	5000	5625
.36	642^{9}	1285^{7}	1929	2571	3214	3857	4500	5143	5786
.37	660^{7}	1321^{4}	1982	2643	3304	3964	4625	5286	5946
.38	678^{6}	1357^{1}	2036	2714	3393	4071	4750	5429	6107
.39	696^{4}	1392^{9}	2089	2786	3482	4179	4875	5571	6268
.40	714^{3}	1428^{6}	2143	2857	3571	4286	5000	5714	6429
.41	732^{1}	1464^{3}	2196	2929	3661	4393	5125	5857	6589
.42	750^{0}	1500^{0}	2250	3000	3750	4500	5250	6000	6750
.43	767^{9}	1535^{7}	2304	3071	3839	4607	5375	6143	6911
.44	785^{7}	1571^{4}	2357	3143	3929	4714	5500	6286	7071
.45	803^{6}	1607^{1}	2411	3214	4018	4821	5625	6429	7232
.46	821^{4}	1642^{9}	2464	3286	4107	4929	5750	6571	7393
.47	839^{3}	1678^{6}	2518	3357	4196	5036	5875	6714	7554
.48	857^{1}	1714^{3}	2571	3429	4286	5143	6000	6857	7714
.49	875^{0}	1750^{0}	2625	3500	4375	5250	6125	7000	7875
.50	892^{9}	1785^{7}	2679	3571	4464	5357	6250	7143	8036
.51	910^{7}	1821^{4}	2732	3643	4554	5464	6375	7286	8196
.52	928^{6}	1857^{1}	2786	3714	4643	5571	6500	7429	8357
.53	946^{4}	1892^{9}	2839	3786	4732	5679	6625	7571	8518
.54	964^{3}	1928^{6}	2893	3857	4821	5786	6756	7714	8679
.55	982^{1}	1964^{3}	2946	3929	4911	5893	6875	7857	8839
.56	1000^{0}	2000^{0}	3000	4000	5000	6000	7000	8000	9000
.57	1017^{9}	2035^{7}	3054	4071	5089	6107	7125	8143	9161
.58	1035^{7}	2071^{4}	3107	4143	5179	6214	7250	8286	9321
.59	1053^{6}	2107^{1}	3161	4214	5268	6321	7375	8429	9482
.60	1071^{4}	2142^{9}	3214	4286	5357	6429	7500	8571	9643
.61	1089^{3}	2178^{6}	3268	4357	5446	6536	7625	8714	9804
.62	1107^{1}	2214^{3}	3321	4429	5536	6643	7750	8857	9964
.63	1125^{0}	2250^{0}	3375	4500	5625	6750	7875	9000	10125
.64	1142^{9}	2285^{7}	3429	4571	5714	6857	8000	9143	10286
.65	1160^{7}	2321^{4}	3482	4643	5804	6964	8125	9286	10446
.66	1178^{6}	2357^{1}	3536	4714	5893	7071	8250	9429	10607

Weight.	1000 100 10	2000 200 20	3000 300 30	4000 400 40	5000 500 50	6000 600 60	7000 700 70	8000 800 80	9000 900 90
.67	1196^{4}	2392^{9}	3589	4786	5982	7179	8375	9571	10768
.68	1214^{3}	2428^{6}	3643	4857	6071	7286	8500	9714	10929
.69	1232^{1}	2464^{3}	3696	4929	6161	7393	8625	9857	11089
.70	1250^{0}	2500^{0}	3750	5000	6250	7500	8750	10000	11250
.71	1267^{9}	2535^{7}	3804	5071	6339	7607	8875	10143	11411
.72	1285^{7}	2571^{4}	3857	5143	6429	7714	9000	10286	11571
.73	1303^{6}	2607^{1}	3911	5214	6518	7821	9125	10429	11732
.74	1321^{4}	2642^{9}	3964	5286	6607	7929	9250	10571	11893
.75	1339^{3}	2678^{6}	4018	5357	6696	8036	9375	10714	12054
.76	1357^{1}	2714^{3}	4071	5429	6786	8143	9500	10857	12214
.77	1375^{0}	2750^{0}	4125	5500	6875	8250	9625	11000	12375
.78	1392^{9}	2785^{7}	4179	5571	6964	8357	9750	11143	12536
.79	1410^{7}	2821^{4}	4232	5643	7054	8464	9875	11286	12696
.80	1428^{6}	2857^{1}	4286	5714	7143	8571	10000	11429	12857
.81	1446^{4}	2892^{9}	4339	5786	7232	8679	10125	11571	13018
.82	1464^{3}	2928^{6}	4393	5857	7321	8786	10250	11714	13179
.83	1482^{1}	2964^{3}	4446	5929	7411	8893	10375	11857	13339
.84	1500^{0}	3000^{0}	4500	6000	7500	9000	10500	12000	13500
.85	1517^{9}	3035^{7}	4554	6071	7589	9107	10625	12143	13661
.86	1535^{7}	3071^{4}	4607	6143	7679	9214	10750	12286	13821
.87	1553^{6}	3107^{1}	4661	6214	7768	9321	10875	12429	13982
.88	1571^{4}	3142^{9}	4714	6286	7857	9429	11000	12571	14143
.89	1589^{3}	3178^{6}	4768	6357	7946	9536	11125	12714	14304
.90	1607^{1}	3214^{3}	4821	6429	8036	9643	11250	12857	14464
.91	1625^{0}	3250^{0}	4875	6500	8125	9750	11375	13000	14625
.92	1642^{9}	3285^{7}	4929	6571	8214	9857	11500	13143	14786
.93	1660^{7}	3321^{4}	4982	6643	8304	9964	11625	13286	14946
.94	1678^{6}	3357^{1}	5036	6714	8393	10071	11750	13429	15107
.95	1696^{4}	3392^{9}	5089	6786	8482	10179	11875	13571	15268
.96	1714^{3}	3428^{6}	5143	6857	8571	10286	12000	13714	15429
.97	1732^{1}	3464^{3}	5196	6929	8661	10393	12125	13857	15589
.98	1750^{0}	3500^{0}	5250	7000	8750	10500	12250	14000	15750
.99	1767^{9}	3535^{7}	5304	7071	8839	10607	12375	14143	15911
1.00	1785^{7}	3571^{4}	5357	7143	8929	10714	12500	14286	16071
1.01	1803^{6}	3607^{1}	5411	7214	9018	10821	12625	14429	16232
1.02	1821^{4}	3642^{9}	5464	7286	9107	10929	12750	14571	16393
1.03	1839^{3}	3678^{6}	5518	7357	9196	11036	12875	14714	16554
1.04	1857^{1}	3714^{3}	5571	7429	9286	11143	13000	14857	16714
1.05	1875^{0}	3750^{0}	5625	7500	9375	11250	13125	15000	16875
1.06	1892^{9}	3785^{7}	5679	7571	9464	11357	13250	15143	17036
1.07	1910^{7}	3821^{4}	5732	7643	9554	11464	13375	15286	17196
1.08	1928^{6}	3857^{1}	5786	7714	9643	11571	13500	15429	17357
1.09	1946^{4}	3892^{9}	5839	7786	9732	11679	13625	15571	17518
1.10	1964^{3}	3928^{6}	5893	7857	9821	11786	13750	15714	17679
1.11	1982^{1}	3964^{3}	5946	7929	9911	11893	13875	15857	17839
1.12	2000^{0}	4000^{0}	6000	8000	10000	12000	14000	16000	18000
1.13	2017^{9}	4035^{7}	6054	8071	10089	12107	14125	16143	18161
1.14	2035^{7}	4071^{4}	6107	8143	10179	12214	14250	16286	18321
1.15	2053^{6}	4107^{1}	6161	8214	10268	12321	14375	16429	18482
1.16	2071^{4}	4142^{9}	6214	8286	10357	12429	14500	16571	18643
1.17	2089^{3}	4178^{6}	6268	8357	10446	12536	14625	16714	18804
1.18	2107^{1}	4214^{3}	6321	8429	10536	12643	14750	16857	18964
1.19	2125^{0}	4250^{0}	6375	8500	10625	12750	14875	17000	19125

Weight.	1000^{0} 100 10	2000^{0} 200 20	3000 300 30	4000 400 40	5000 500 50	6000 600 60	7000 700 70	8000 800 80	9000 900 90
Bush.	2857^{1}	5714^{3}	8571	11429	14286	17143	20000	22857	25714
¼	7^{1}	14^{3}	21	29	36	43	50	57	64
⅓	9^{5}	19^{0}	29	38	48	57	67	76	86
½	14^{3}	28^{6}	43	57	71	86	100	114	129
⅔	19^{0}	38^{1}	57	76	95	114	133	152	171
¾	21^{4}	42^{9}	64	86	107	129	150	171	193
Price per Bu.									
.17	485^{7}	971^{4}	1457	1943	2429	2914	3400	3886	4371
.18	514^{3}	1028^{6}	1543	2057	2571	3086	3600	4114	4629
.19	542^{9}	1085^{7}	1629	2171	2714	3257	3800	4343	4886
.20	571^{4}	1142^{9}	1714	2286	2857	3429	4000	4571	5143
.21	600^{0}	1200^{0}	1800	2400	3000	3600	4200	4800	5400
.22	628^{6}	1257^{1}	1886	2514	3143	3771	4400	5029	5657
.23	657^{1}	1314^{3}	1971	2629	3286	3943	4600	5257	5914
.24	685^{7}	1371^{4}	2057	2743	3429	4114	4800	5486	6171
.25	714^{3}	1428^{6}	2143	2857	3571	4286	5000	5714	6429
.26	742^{9}	1485^{7}	2229	2971	3714	4457	5200	5943	6686
.27	771^{4}	1542^{9}	2314	3086	3857	4629	5400	6171	6943
.28	800^{0}	1600^{0}	2400	3200	4000	4800	5600	6400	7200
.29	828^{6}	1657^{1}	2486	3314	4143	4971	5800	6629	7457
.30	857^{1}	1714^{3}	2571	3429	4286	5143	6000	6857	7714
.31	885^{7}	1771^{4}	2657	3543	4429	5314	6200	7086	7971
.32	914^{3}	1828^{6}	2743	3657	4571	5486	6400	7314	8229
.33	942^{9}	1885^{7}	2829	3771	4714	5657	6600	7543	8486
.34	971^{4}	1942^{9}	2914	3886	4857	5829	6800	7771	8743
.35	1000^{0}	2000^{0}	3000	4000	5000	6000	7000	8000	9000
.36	1028^{6}	2057^{1}	3086	4114	5143	6171	7200	8229	9257
.37	1057^{1}	2114^{3}	3171	4229	5286	6343	7400	8457	9514
.38	1085^{7}	2171^{4}	3257	4343	5429	6514	7600	8686	9771
.39	1114^{3}	2228^{6}	3343	4457	5571	6686	7800	8914	10029
.40	1142^{9}	2285^{7}	3429	4571	5714	6857	8000	9143	10286
.41	1171^{4}	2342^{9}	3514	4686	5857	7029	8200	9371	10543
.42	1200^{0}	2400^{0}	3600	4800	6000	7200	8400	9600	10800
.43	1228^{6}	2457^{1}	3686	4914	6143	7371	8600	9829	11057
.44	1257^{1}	2514^{3}	3771	5029	6286	7543	8800	10057	11314
.45	1285^{7}	2571^{4}	3857	5143	6429	7714	9000	10286	11571
.46	1314^{3}	2628^{6}	3943	5257	6571	7886	9200	10514	11829
.47	1342^{9}	2685^{7}	4029	5371	6714	8057	9400	10743	12086
.48	1371^{4}	2742^{9}	4114	5486	6857	8229	9600	10971	12343
.49	1400^{0}	2800^{0}	4200	5600	7000	8400	9800	11200	12600
.50	1428^{6}	2857^{1}	4286	5714	7143	8571	10000	11429	12857
.51	1457^{1}	2914^{3}	4371	5829	7286	8743	10200	11657	13114
.52	1485^{7}	2971^{4}	4457	5943	7429	8914	10400	11886	13371
.53	1514^{3}	3028^{6}	4543	6057	7571	9086	10600	12114	13629
.54	1542^{9}	3085^{7}	4629	6171	7714	9257	10800	12343	13886
.55	1571^{4}	3142^{9}	4714	6286	7857	9429	11000	12571	14143
.56	1600^{0}	3200^{0}	4800	6400	8000	9600	11200	12800	14400
.57	1628^{6}	3257^{1}	4886	6514	8143	9771	11400	13029	14657
.58	1657^{1}	3314^{3}	4971	6629	8286	9943	11600	13257	14914
.59	1685^{7}	3371^{4}	5057	6743	8429	10114	11800	13486	15171
.60	1714^{3}	3428^{6}	5143	6857	8571	10286	12000	13714	15429
.61	1742^{9}	3485^{7}	5229	6971	8714	10457	12200	13943	15686
.62	1771^{4}	3542^{9}	5314	7086	8857	10629	12400	14171	15943
.63	1800^{0}	3600^{0}	5400	7200	9000	10800	12600	14400	16200

Weight.	1000^0 100 10	2000^0 200 20	3000 300 30	4000 400 40	5000 500 50	6000 600 60	7000 700 70	8000 800 80	9000 900 90
Bush.	3125^0	6250^0	9375	12500	15625	18750	21875	25000	28125
Price per Bu. ¼	7^8	15^6	23	31	39	47	55	63	70
⅓	10^4	20^8	31	42	52	63	73	83	94
½	15^6	31^2	47	63	78	94	109	125	141
⅔	20^8	41^7	63	83	104	125	146	167	188
¾	23^4	46^9	70	94	117	141	164	188	211
.18	562^5	1125^0	1688	2250	2813	3375	3938	4500	5063
.19	593^8	1187^5	1781	2375	2969	3563	4156	4750	5344
.20	625^0	1250^0	1875	2500	3125	3750	4375	5000	5625
.21	656^3	1312^5	1969	2625	3281	3938	4594	5250	5906
.22	687^5	1375^0	2063	2750	3438	4125	4813	5500	6188
.23	718^8	1437^5	2156	2875	3594	4313	5031	5750	6469
.24	750^0	1500^0	2250	3000	3750	4500	5250	6000	6750
.25	781^3	1562^5	2344	3125	3906	4688	5469	6250	7031
.26	812^5	1625^0	2438	3250	4063	4875	5688	6500	7313
.27	843^8	1687^5	2531	3375	4219	5063	5906	6750	7594
.28	875^0	1750^0	2625	3500	4375	5250	6125	7000	7875
.29	906^3	1812^5	2719	3625	4531	5438	6344	7250	8156
.30	937^5	1875^0	2813	3750	4688	5625	6563	7500	8438
.31	968^8	1937^5	2906	3875	4844	5813	6781	7750	8719
.32	1000^0	2000^0	3000	4000	5000	6000	7000	8000	9000
.33	1031^3	2062^5	3094	4125	5156	6188	7219	8250	9281
.34	1062^5	2125^0	3188	4250	5313	6375	7438	8500	9563
.35	1093^8	2187^5	3281	4375	5469	6563	7656	8750	9844
.36	1125^0	2250^0	3375	4500	5625	6750	7875	9000	10125
.37	1156^3	2312^5	3469	4625	5781	6938	8094	9250	10406
.38	1187^5	2375^0	3563	4750	5938	7125	8313	9500	10688
.39	1218^8	2437^5	3656	4875	6094	7313	8531	9750	10969
.40	1250^0	2500^0	3750	5000	6250	7500	8750	10000	11250
.41	1281^3	2562^5	3844	5125	6406	7688	8969	10250	11531
.42	1312^5	2625^0	3938	5250	6563	7875	9188	10500	11813
.43	1343^8	2687^5	4031	5375	6719	8063	9406	10750	12094
.44	1375^0	2750^0	4125	5500	6875	8250	9625	11000	12375
.45	1406^3	2812^5	4219	5625	7031	8438	9844	11250	12656
.46	1437^5	2875^0	4313	5750	7188	8625	10063	11500	12938
.47	1468^8	2937^5	4406	5875	7344	8813	10281	11750	13219
.48	1500^0	3000^0	4500	6000	7500	9000	10500	12000	13500
.49	1531^3	3062^5	4594	6125	7656	9188	10719	12250	13781
.50	1562^5	3125^0	4688	6250	7813	9375	10938	12500	14063
.51	1593^8	3187^5	4781	6375	7969	9563	11156	12750	14344
.52	1625^0	3250^0	4875	6500	8125	9750	11375	13000	14625
.53	1656^3	3312^5	4969	6625	8281	9938	11594	13250	14906
.54	1687^5	3375^0	5063	6750	8438	10125	11813	13500	15188
.55	1718^8	3437^5	5156	6875	8594	10313	12031	13750	15469
.56	1750^0	3500^0	5250	7000	8750	10500	12250	14000	15750
.57	1781^3	3562^5	5344	7125	8906	10688	12469	14250	16031
.58	1812^5	3625^0	5438	7250	9063	10875	12688	14500	16313
.59	1843^8	3687^5	5531	7375	9219	11063	12906	14750	16594
.60	1875^0	3750^0	5625	7500	9375	11250	13125	15000	16875
.61	1906^3	3812^5	5719	7625	9531	11438	13344	15250	17156
.62	1937^5	3875^0	5813	7750	9688	11625	13563	15500	17438
.63	1968^8	3937^5	5906	7875	9844	11813	13781	15750	17719
.64	2000^0	4000^0	6000	8000	10000	12000	14000	16000	18000

Weight	1000 / 100 / 10	2000 / 200 / 20	3000 / 300 / 30	4000 / 400 / 40	5000 / 500 / 50	6000 / 600 / 60	7000 / 700 / 70	8000 / 800 / 80	9000 / 900 / 90
Bush.	2083^{3}	4166^{7}	6250	8333	10417	12500	14583	16667	18750
Price per Bu. 1/4	5^{2}	10^{4}	16	21	26	31	36	42	47
1/3	6^{9}	13^{9}	21	28	35	42	49	56	63
1/2	10^{4}	20^{8}	31	42	52	63	73	83	94
2/3	13^{9}	27^{8}	42	56	69	83	97	111	125
3/4	15^{6}	31^{3}	47	63	78	94	109	125	141
.45	937^{5}	1875^{0}	2813	3750	4688	5625	6563	7500	8438
.46	958^{3}	1916^{7}	2875	3833	4792	5750	6708	7667	8625
.47	979^{2}	1958^{3}	2938	3917	4896	5875	6854	7833	8813
.48	1000^{0}	2000^{0}	3000	4000	5000	6000	7000	8000	9000
.49	1020^{8}	2041^{7}	3063	4083	5104	6125	7146	8167	9188
.50	1041^{7}	2083^{3}	3125	4167	5208	6250	7292	8333	9375
.51	1062^{5}	2125^{0}	3188	4250	5313	6375	7438	8500	9563
.52	1083^{3}	2166^{7}	3250	4333	5417	6500	7583	8667	9750
.53	1104^{2}	2208^{3}	3313	4417	5521	6625	7729	8833	9938
.54	1125^{0}	2250^{0}	3375	4500	5625	6750	7875	9000	10125
.55	1145^{8}	2291^{7}	3438	4583	5729	6875	8021	9167	10313
.56	1166^{7}	2333^{3}	3500	4667	5833	7000	8167	9333	10500
.57	1187^{5}	2375^{0}	3563	4750	5938	7125	8313	9500	10688
.58	1208^{3}	2416^{7}	3625	4833	6042	7250	8458	9667	10875
.59	1229^{2}	2458^{3}	3688	4917	6146	7375	8604	9833	11063
.60	1250^{0}	2500^{0}	3750	5000	6250	7500	8750	10000	11250
.61	1270^{8}	2541^{7}	3813	5083	6354	7625	8896	10167	11438
.62	1291^{7}	2583^{3}	3875	5167	6458	7750	9042	10333	11625
.63	1312^{5}	2625^{0}	3938	5250	6563	7875	9188	10500	11813
.64	1333^{3}	2666^{7}	4000	5333	6667	8000	9333	10667	12000
.65	1354^{2}	2708^{3}	4063	5417	6771	8125	9479	10833	12188
.66	1375^{0}	2750^{0}	4125	5500	6875	8250	9625	11000	12375
.67	1395^{8}	2791^{7}	4188	5583	6979	8375	9771	11167	12563
.68	1416^{7}	2833^{3}	4250	5667	7083	8500	9917	11333	12750
.69	1437^{5}	2875^{0}	4313	5750	7188	8625	10063	11500	12938
.70	1458^{3}	2916^{7}	4375	5833	7292	8750	10208	11667	13125
.71	1479^{2}	2958^{3}	4438	5917	7396	8875	10354	11833	13313
.72	1500^{0}	3000^{0}	4500	6000	7500	9000	10500	12000	13500
.73	1520^{8}	3041^{7}	4563	6083	7604	9125	10642	12167	13688
.74	1541^{7}	3083^{3}	4625	6167	7708	9250	10792	12333	13875
.75	1562^{5}	3125^{0}	4688	6250	7813	9375	10938	12500	14063
.76	1583^{3}	3166^{7}	4750	6333	7917	9500	11083	12667	14250
.77	1604^{2}	3208^{3}	4813	6417	8021	9625	11229	12833	14438
.78	1625^{0}	3250^{0}	4875	6500	8125	9750	11375	13000	14625
.79	1645^{8}	3291^{7}	4938	6583	8229	9875	11521	13167	14813
.80	1666^{7}	3333^{3}	5000	6667	8333	10000	11667	13333	15000
.81	1687^{5}	3375^{0}	5063	6750	8438	10125	11813	13500	15188
.82	1708^{3}	3416^{7}	5125	6833	8542	10250	11958	13667	15375
.83	1729^{2}	3458^{3}	5188	6917	8646	10375	12104	13833	15563
.84	1750^{0}	3500^{0}	5250	7000	8750	10500	12250	14000	15750
.85	1770^{8}	3541^{7}	5313	7083	8854	10625	12396	14167	15938
.86	1791^{7}	3583^{3}	5375	7167	8958	10750	12542	14333	16125
.87	1812^{5}	3625^{0}	5438	7250	9063	10875	12688	14500	16313
.88	1833^{3}	3666^{7}	5500	7333	9167	11000	12833	14667	16500
.89	1854^{2}	3708^{3}	5563	7417	9271	11125	12979	14833	16688
.90	1875^{0}	3750^{0}	5625	7500	9375	11250	13125	15000	16875
.91	1895^{8}	3791^{7}	5688	7583	9479	11375	13271	15167	17063

Weight.	1000^0 100 10	2000^0 200 20	3000 300 30	4000 400 40	5000 500 50	6000 600 60	7000 700 70	8000 800 80	9000 900 90
Bush.	1428^6	2857^1	4286	5714	7143	8571	10000	11429	12857
½	7^1	14^3	21	29	36	43	50	57	64
Price per Bu. .20	285^7	571^4	857	1143	1429	1714	2000	2286	2571
.21	300^0	600^0	900	1200	1500	1800	2100	2400	2700
.22	314^3	628^6	943	1257	1571	1886	2200	2514	2829
.23	328^6	657^1	986	1314	1643	1971	2300	2629	2957
.24	342^9	685^7	1029	1371	1714	2057	2400	2743	3086
.25	357^1	714^3	1071	1429	1786	2143	2500	2857	3214
.26	371^4	742^9	1114	1486	1857	2229	2600	2971	3343
.27	385^7	771^4	1157	1543	1929	2314	2700	3086	3471
.28	400^0	800^0	1200	1600	2000	2400	2800	3200	3600
.29	414^3	828^6	1243	1657	2071	2486	2900	3314	3729
.30	428^6	857^1	1286	1714	2143	2571	3000	3429	3857
.31	442^9	885^7	1329	1771	2214	2657	3100	3543	3986
.32	457^1	914^3	1371	1829	2286	2743	3200	3657	4114
.33	471^4	942^9	1414	1886	2357	2829	3300	3771	4243
.34	485^7	971^4	1457	1943	2429	2914	3400	3886	4371
.35	500^0	1000^0	1500	2000	2500	3000	3500	4000	4500
.36	514^3	1028^6	1543	2057	2571	3086	3600	4114	4629
.37	528^6	1057^1	1586	2114	2643	3171	3700	4229	4757
.38	542^9	1085^7	1629	2171	2714	3257	3800	4343	4886
.39	557^1	1114^3	1671	2229	2786	3343	3900	4457	5014
.40	571^4	1142^9	1714	2286	2857	3429	4000	4571	5143
.41	585^7	1171^4	1757	2343	2929	3514	4100	4686	5271
.42	600^0	1200^0	1800	2400	3000	3600	4200	4800	5400
.43	614^3	1228^6	1843	2457	3071	3686	4300	4914	5529
.44	628^6	1257^1	1886	2514	3143	3771	4400	5029	5657
.45	642^9	1285^7	1929	2571	3214	3857	4500	5143	5786
.46	657^1	1314^3	1971	2629	3286	3943	4600	5257	5914
.47	671^4	1342^9	2014	2686	3357	4029	4700	5371	6043
.48	685^7	1371^4	2057	2743	3429	4114	4800	5486	6171
.49	700^0	1400^0	2100	2800	3500	4200	4900	5600	6300
.50	714^3	1428^6	2143	2857	3571	4286	5000	5714	6429
.51	728^6	1457^1	2186	2914	3643	4371	5100	5829	6557
.52	742^9	1485^7	2229	2971	3714	4457	5200	5943	6686
.53	757^1	1514^3	2271	3029	3786	4543	5300	6057	6814
.54	771^4	1542^9	2314	3086	3857	4629	5400	6171	6943
.55	785^7	1571^4	2357	3143	3929	4714	5500	6286	7071
.56	800^0	1600^0	2400	3200	4000	4800	5600	6400	7200
.57	814^3	1628^6	2443	3257	4071	4886	5700	6514	7329
.58	828^6	1657^1	2486	3314	4143	4971	5800	6629	7457
.59	842^9	1685^7	2529	3371	4214	5057	5900	6743	7586
.60	857^1	1714^3	2571	3429	4286	5143	6000	6857	7714
.61	871^4	1742^9	2614	3486	4357	5229	6100	6971	7843
.62	885^7	1771^4	2657	3543	4429	5314	6200	7086	7971
.63	900^0	1800^0	2700	3600	4500	5400	6300	7200	8100
.64	914^3	1828^6	2743	3657	4571	5486	6400	7314	8229
.65	928^6	1857^1	2786	3714	4643	5571	6500	7429	8357
.66	942^9	1885^7	2829	3771	4714	5657	6600	7543	8486
.67	957^1	1914^3	2871	3829	4786	5743	6700	7657	8614
.68	971^4	1942^9	2914	3886	4857	5829	6800	7771	8743
.69	985^7	1971^4	2957	3943	4929	5914	6900	7886	8871
.70	1000^0	2000^0	3000	4000	5000	6000	7000	8000	9000

Table Showing Value of CORN in Ear—75 & 80 lbs. to Bu.

Weight.	1000" 100 10	2000" 200 20	3000 300 30	4000 400 40	5000 500 50	6000 600 60	7000 700 70	8000 800 80	9000 900 90
Bush.	1333^{3}	2666^{7}	4000	5333	6667	8000	9333	10667	12000
½	6^{7}	13^{3}	20	27	33	40	47	53	60
Price per Bush., 75 lbs.									
.21	280^{0}	560^{0}	840	1120	1400	1680	1960	2240	2520
.22	293^{3}	586^{7}	880	1173	1467	1760	2053	2347	2640
.23	306^{7}	613^{3}	920	1227	1533	1840	2147	2453	2760
.24	320^{0}	640^{0}	960	1280	1600	1920	2240	2560	2880
.25	333^{3}	666^{7}	1000	1333	1667	2000	2333	2667	3000
.26	346^{7}	693^{3}	1040	1387	1733	2080	2427	2773	3120
.27	360^{0}	720^{0}	1080	1440	1800	2160	2520	2880	3240
.28	373^{3}	746^{7}	1120	1493	1867	2240	2613	2987	3360
.29	386^{7}	773^{3}	1160	1547	1933	2320	2707	3093	3480
.30	400^{0}	800^{0}	1200	1600	2000	2400	2800	3200	3600
.31	413^{3}	826^{7}	1240	1653	2067	2480	2893	3307	3720
.32	426^{7}	853^{3}	1280	1707	2133	2560	2987	3413	3840
.33	440^{0}	880^{0}	1320	1760	2200	2640	3080	3520	3960
.34	453^{3}	906^{7}	1360	1813	2267	2720	3173	3627	4080
.35	466^{7}	933^{3}	1400	1867	2333	2800	3267	3733	4200
.36	480^{0}	960^{0}	1440	1920	2400	2880	3360	3840	4320
.37	493^{3}	986^{7}	1480	1973	2467	2960	3453	3947	4440
.38	506^{7}	1013^{3}	1520	2027	2533	3040	3547	4053	4560
.39	520^{0}	1040^{0}	1560	2080	2600	3120	3640	4160	4680
.40	533^{3}	1066^{7}	1600	2133	2667	3200	3733	4267	4800
.41	546^{7}	1093^{3}	1640	2187	2733	3280	3827	4373	4920
.42	560^{0}	1120^{0}	1680	2240	2800	3360	3920	4480	5040
.43	573^{3}	1146^{7}	1720	2293	2867	3440	4013	4587	5160
.44	586^{7}	1173^{3}	1760	2347	2933	3520	4107	4693	5280
.45	600^{0}	1200^{0}	1800	2400	3000	3600	4200	4800	5400
46	613^{3}	1226^{7}	1840	2453	3067	3680	4293	4907	5520
.47	626^{7}	1253^{3}	1880	2507	3133	3760	4387	5013	5640
.48	640^{0}	1280^{0}	1920	2560	3200	3840	4480	5120	5760
49	653^{3}	1306^{7}	1960	2613	3267	3920	4573	5227	5880
.50	666^{7}	1333^{3}	2000	2667	3333	4000	4667	5333	6000
Bush.	1250^{0}	2500^{0}	3750	5000	6250	7500	8750	10000	11250
½	6^{3}	12^{5}	19	25	31	38	44	50	56
Price per Bush., 80 lbs.									
.22	275^{0}	550^{0}	825	1100	1375	1650	1925	2200	2475
.23	287^{5}	575^{0}	863	1150	1438	1725	2013	2300	2588
.24	300^{0}	600^{0}	900	1200	1500	1800	2100	2400	2700
.25	312^{5}	625^{0}	938	1250	1563	1875	2188	2500	2813
.26	325^{0}	650^{0}	975	1300	1625	1950	2275	2600	2925
.27	337^{5}	675^{0}	1013	1350	1688	2025	2363	2700	3038
.28	350^{0}	700^{0}	1050	1400	1750	2100	2450	2800	3150
.29	362^{5}	725^{0}	1088	1450	1813	2175	2538	2900	3263
.30	375^{0}	750^{0}	1125	1500	1875	2250	2625	3000	3375
.31	387^{5}	775^{0}	1163	1550	1938	2325	2713	3100	3488
.32	400^{0}	800^{0}	1200	1600	2000	2400	2800	3200	3600
.33	412^{5}	825^{0}	1238	1650	2063	2475	2888	3300	3713
.34	425^{0}	850^{0}	1275	1700	2125	2550	2975	3400	3825
.35	437^{5}	875^{0}	1313	1750	2188	2625	3063	3500	3938
.36	450^{0}	900^{0}	1350	1800	2250	2700	3150	3600	4050
.37	462^{5}	925^{0}	1388	1850	2313	2775	3238	3700	4163
.38	475^{0}	950^{0}	1425	1900	2375	2850	3325	3800	4275
.39	487^{5}	975^{0}	1463	1950	2438	2925	3413	3900	4388
.40	500^{0}	1000^{0}	1500	2000	2500	3000	3500	4000	4500

Weight / Price per Ton	1000^0 100 10	2000^0 200 20	3000 300 30	4000 400 40	5000 500 50	6000 600 60	7000 700 70	8000 800 80	9000 900 90
5	2^5	5^0	8	10	13	15	18	20	23
.10	5^0	10^0	15	20	25	30	35	40	45
.25	12^5	25^0	38	50	63	75	88	100	113
.50	25^0	50^0	75	100	125	150	175	200	225
2.00	100^0	200^0	300	400	500	600	700	800	900
2.25	112^5	225^0	338	450	563	675	788	900	1013
2.50	125^0	250^0	375	500	625	750	875	1000	1125
2.75	137^5	275^0	413	550	688	825	963	1100	1238
3.00	150^0	300^0	450	600	750	900	1050	1200	1350
3.25	162^5	325^0	488	650	813	975	1138	1300	1463
3.50	175^0	350^0	525	700	875	1050	1225	1400	1575
3.75	187^5	375^0	563	750	938	1125	1313	1500	1688
4.00	200^0	400^0	600	800	1000	1200	1400	1600	1800
4.25	212^5	425^0	638	850	1063	1275	1488	1700	1913
4.50	225^0	450^0	675	900	1125	1350	1575	1800	2025
4.75	237^5	475^0	713	950	1188	1425	1663	1900	2138
5.00	250^0	500^0	750	1000	1250	1500	1750	2000	2250
5.25	262^5	525^0	788	1050	1313	1575	1838	2100	2363
5.50	275^0	550^0	825	1100	1375	1650	1925	2200	2475
5.75	287^5	575^0	863	1150	1438	1725	2013	2300	2588
6.00	300^0	600^0	900	1200	1500	1800	2100	2400	2700
6.25	312^5	625^0	938	1250	1563	1875	2188	2500	2813
6.50	325^0	650^0	975	1300	1625	1950	2275	2600	2925
6.75	337^5	675^0	1013	1350	1688	2025	2363	2700	3038
7.00	350^0	700^0	1050	1400	1750	2100	2450	2800	3150
7.25	362^5	725^0	1088	1450	1813	2175	2538	2900	3263
7.50	375^0	750^0	1125	1500	1875	2250	2625	3000	3375
8.00	400^0	800^0	1200	1600	2000	2400	2800	3200	3600
8.50	425^0	850^0	1275	1700	2125	2550	2975	3400	3825
9.00	450^0	900^0	1350	1800	2250	2700	3150	3600	4050
9.50	475^0	950^0	1425	1900	2375	2850	3325	3800	4275
10.00	500^0	1000^0	1500	2000	2500	3000	3500	4000	4500
10.50	525^0	1050^0	1575	2100	2625	3150	3675	4200	4725
11.00	550^0	1100^0	1650	2200	2750	3300	3850	4400	4950
11.50	575^0	1150^0	1725	2300	2875	3450	4025	4600	5175
12.00	600^0	1200^0	1800	2400	3000	3600	4200	4800	5400
12.50	625^0	1250^0	1875	2500	3125	3750	4375	5000	5625
13.00	650^0	1300^0	1950	2600	3250	3900	4550	5200	5850
13.50	675^0	1350^0	2025	2700	3375	4050	4725	5400	6075
14.00	700^0	1400^0	2100	2800	3500	4200	4900	5600	6300
14.50	725^0	1450^0	2175	2900	3625	4350	5075	5800	6525
15.00	750^0	1500^0	2250	3000	3750	4500	5250	6000	6750
16.00	800^0	1600^0	2400	3200	4000	4800	5600	6400	7200
17.00	850^0	1700^0	2550	3400	4250	5100	5950	6800	7650
18.00	900^0	1800^0	2700	3600	4500	5400	6300	7200	8100
19.00	950^0	1900^0	2850	3800	4750	5700	6650	7600	8550
20.00	1000^0	2000^0	3000	4000	5000	6000	7000	8000	9000
21.00	1050^0	2100^0	3150	4200	5250	6300	7350	8400	9450
22.00	1100^0	2200^0	3300	4400	5500	6600	7700	8800	9900
23.00	1150^0	2300^0	3450	4600	5750	6900	8050	9200	10350
24.00	1200^0	2400^0	3600	4800	6000	7200	8400	9600	10800
25.00	1250^0	2500^0	3750	5000	6250	7500	8750	10000	11250
26.00	1300^0	2600^0	3900	5200	6500	7800	9100	10400	11700

Table showing Value of articles sold by the 1000; Lumber, &c.

Quan'ty	1000	2000	3000	4000	5000	6000	7000	8000	9000
	100	200	300	400	500	600	700	800	900
Price ⅌ M.	10	20	30	40	50	60	70	80	90
5	5^{0}	10^{0}	15	20	25	30	35	40	45
.10	10^{0}	20^{0}	30	40	50	60	70	80	90
.25	25^{0}	50^{0}	75	100	125	150	175	200	225
.50	50^{0}	100^{0}	150	200	250	300	350	400	450
2.00	200^{0}	400^{0}	600	800	1000	1200	1400	1600	1800
2.25	225^{0}	450^{0}	675	900	1125	1350	1575	1800	2025
2.50	250^{0}	500^{0}	750	1000	1250	1500	1750	2000	2250
2.75	275^{0}	550^{0}	825	1100	1375	1650	1925	2200	2475
3.00	300^{0}	600^{0}	900	1200	1500	1800	2100	2400	2700
3.25	325^{0}	650^{0}	975	1300	1625	1950	2275	2600	2925
3.50	350^{0}	700^{0}	1050	1400	1750	2100	2450	2800	3150
3.75	375^{0}	750^{0}	1125	1500	1875	2250	2625	3000	3375
4.00	400^{0}	800^{0}	1200	1600	2000	2400	2800	3200	3600
4.25	425^{0}	850^{0}	1275	1700	2125	2550	2975	3400	3825
4.50	450^{0}	900^{0}	1350	1800	2250	2700	3150	3600	4050
4.75	475^{0}	950^{0}	1425	1900	2375	2850	3325	3800	4275
5.00	500^{0}	1000^{0}	1500	2000	2500	3000	3500	4000	4500
5.50	550^{0}	1100^{0}	1650	2200	2750	3300	3850	4400	4950
6.00	600^{0}	1200^{0}	1800	2400	3000	3600	4200	4800	5400
6.50	650^{0}	1300^{0}	1950	2600	3250	3900	4550	5200	5850
7.00	700^{0}	1400^{0}	2100	2800	3500	4200	4900	5600	6300
7.50	750^{0}	1500^{0}	2250	3000	3750	4500	5250	6000	6750
8.00	800^{0}	1600^{0}	2400	3200	4000	4800	5600	6400	7200
8.50	850^{0}	1700^{0}	2550	3400	4250	5100	5950	6800	7650
9.00	900^{0}	1800^{0}	2700	3600	4500	5400	6300	7200	8100
9.50	950^{0}	1900^{0}	2850	3800	4750	5700	665[illegible]	7600	8550
10.00	1000^{0}	2000^{0}	3000	4000	5000	6000	7000	8000	9000
11.00	1100^{0}	2200^{0}	3300	4400	5500	6600	7700	8800	9900
12.00	1200^{0}	2400^{0}	3600	4800	6000	7200	8400	9600	10800
13.00	1300^{0}	2600^{0}	3900	5200	6500	7800	9100	10400	11700
14.00	1400^{0}	2800^{0}	4200	5600	7000	8400	9800	11200	12600
15.00	1500^{0}	3000^{0}	4500	6000	7500	9000	10500	12000	13500
16.00	1600^{0}	3200^{0}	4800	6400	8000	9600	11200	12800	14400
17.00	1700^{0}	3400^{0}	5100	6800	8500	10200	11900	13600	15300
18.00	1800^{0}	3600^{0}	5400	7200	9000	10800	12600	14400	16200
19.00	1900^{0}	3800^{0}	5700	7600	9500	11400	13300	15200	17100
20.00	2000^{0}	4000^{0}	6000	8000	10000	12000	14000	16000	18000
21.00	2100^{0}	4200^{0}	6300	8400	10500	12600	14700	16800	18900
22.00	2200^{0}	4400^{0}	6600	8800	11000	13200	15400	17600	19800
23.00	2300^{0}	4600^{0}	6900	9200	11500	13800	16100	18400	20700
24.00	2400^{0}	4800^{0}	7200	9600	12000	14400	16800	19200	21600
25.00	2500^{0}	5000^{0}	7500	10000	12500	15000	17500	20000	22500
26.00	2600^{0}	5200^{0}	7800	10400	13000	15600	18200	20800	23400
27.00	2700^{0}	5400^{0}	8100	10800	13500	16200	18900	21600	24300
28.00	2800^{0}	5600^{0}	8400	11200	14000	16800	19600	22400	25200
29.00	2900^{0}	5800^{0}	8700	11600	14500	17400	20300	23200	26100
30.00	3000^{0}	6000^{0}	9000	12000	15000	18000	21000	24000	27000
31.00	3100^{0}	6200^{0}	9300	12400	15500	18600	21700	24800	27900
32.00	3200^{0}	6400^{0}	9600	12800	16000	19200	22400	25600	28800
33.00	3300^{0}	6600^{0}	9900	13200	16500	19800	23100	26400	29700
34.00	3400^{0}	6800^{0}	10200	13600	17000	20400	23800	27200	30600
35.00	3500^{0}	7000^{0}	10500	14000	17500	21000	24500	28000	31500
36.00	3600^{0}	7200^{0}	10800	14400	18000	21600	25200	28800	32400

Weight.	1000	2000	3000	4000	5000	6000	7000	8000	9000
	100	200	300	400	500	600	700	800	900
Price per Hundred.	10	20	30	40	50	60	70	80	90
5	500	1000	150	200	250	300	350	400	450
.10	1000	2000	300	400	500	600	700	800	900
.15	1500	3000	450	600	750	900	1050	1200	1350
.20	2000	4000	600	800	1000	1200	1400	1600	1800
.25	2500	5000	750	1000	1250	1500	1750	2000	2250
.30	3000	6000	900	1200	1500	1800	2100	2400	2700
.40	4000	8000	1200	1600	2000	2400	2800	3200	3600
.50	5000	10000	1500	2000	2500	3000	3500	4000	4500
.60	6000	12000	1800	2400	3000	3600	4200	4800	5400
.70	7000	14000	2100	2800	3500	4200	4900	5600	6300
.80	8000	16000	2400	3200	4000	4800	5600	6400	7200
.90	9000	18000	2700	3600	4500	5400	6300	7200	8100
1.00	10000	20000	3000	4000	5000	6000	7000	8000	9000
1.25	12500	25000	3750	5000	6250	7500	8750	10000	11250
1.50	15000	30000	4500	6000	7500	9000	10500	12000	13500
1.75	17500	35000	5250	7000	8750	10500	12250	14000	15750
2.00	20000	40000	6000	8000	10000	12000	14000	16000	18000
2.25	22500	45000	6750	9000	11250	13500	15750	18000	20250
2.50	25000	50000	7500	10000	12500	15000	17500	20000	22500
2.75	27500	55000	8250	11000	13750	16500	19250	22000	24750
3.00	30000	60000	9000	12000	15000	18000	21000	24000	27000
3.25	32500	65000	9750	13000	16250	19500	22750	26000	29250
3.50	35000	70000	10500	14000	17500	21000	24500	28000	31500
3.75	37500	75000	11250	15000	18750	22500	26250	30000	33750
4.00	40000	80000	12000	16000	20000	24000	28000	32000	36000
4.25	42500	85000	12750	17000	21250	25500	29750	34000	38250
4.50	45000	90000	13500	18000	22500	27000	31500	36000	40500
4.75	47500	95000	14250	19000	23750	28500	33250	38000	42750
5.00	50000	100000	15000	20000	25000	30000	35000	40000	45000
5.25	52500	105000	15750	21000	26250	31500	36750	42000	47250
5.50	55000	110000	16500	22000	27500	33000	38500	44000	49500
5.75	57500	115000	17250	23000	28750	34500	40250	46000	51750
6.00	60000	120000	18000	24000	30000	36000	42000	48000	54000
6.25	62500	125000	18750	25000	31250	37500	43750	50000	56250
6.50	65000	130000	19500	26000	32500	39000	45500	52000	58500
6.75	67500	135000	20250	27000	33750	40500	47250	54000	60750
7.00	70000	140000	21000	28000	35000	42000	49000	56000	63000
7.25	72500	145000	21750	29000	36250	43500	50750	58000	65250
7.50	75000	150000	22500	30000	37500	45000	52500	60000	67500
7.75	77500	155000	23250	31000	38750	46500	54250	62000	69750
8.00	80000	160000	24000	32000	40000	48000	56000	64000	72000
8.25	82500	165000	24750	33000	41250	49500	57750	66000	74250
8.50	85000	170000	25500	34000	42500	51000	59500	68000	76500
8.75	87500	175000	26250	35000	43750	52500	61250	70000	78750
9.00	90000	180000	27000	36000	45000	54000	63000	72000	81000
9.25	92500	185000	27750	37000	46250	55500	64750	74000	83250
9.50	95000	190000	28500	38000	47500	57000	66500	76000	85500
9.75	97500	195000	29250	39000	48750	58500	68250	78000	87750
10.00	100000	200000	30000	40000	50000	60000	70000	80000	90000
10.25	102500	205000	30750	41000	51250	61500	71750	82000	92250
10.50	105000	210000	31500	42000	52500	63000	73500	84000	94500
10.75	107500	215000	32250	43000	53750	64500	75250	86000	96750
11.00	110000	220000	33000	44000	55000	66000	77000	88000	99000

Princip'l		$100^{0} 10 1	$200^{0} 20 2	$300 30 3	$400 40 4	$500 50 5	$600 60 6	$700 70 7	$800 80 8	$900 90 9
Days.	1	1^{7}	3^{3}	5	7	8	10	12	13	15
	2	3^{3}	6^{7}	10	13	17	20	23	27	30
	3	5^{0}	10^{0}	15	20	25	30	35	40	45
	4	6^{7}	13^{3}	20	27	33	40	47	53	60
	5	8^{3}	16^{7}	25	33	42	50	58	67	75
	6	10^{0}	20^{0}	30	40	50	60	70	80	90
	7	11^{7}	23^{3}	35	47	58	70	82	93	105
	8	13^{3}	26^{7}	40	53	67	80	93	107	120
	9	15^{0}	30^{0}	45	60	75	90	105	120	135
	10	16^{7}	33^{3}	50	67	83	100	117	133	150
	11	18^{3}	36^{7}	55	73	92	110	128	147	165
	12	20^{0}	40^{0}	60	80	100	120	140	160	180
	13	21^{7}	43^{3}	65	87	108	130	152	173	195
	14	23^{3}	46^{7}	70	93	117	140	163	187	210
	15	25^{0}	50^{0}	75	100	125	150	175	200	225
	16	26^{7}	53^{3}	80	107	133	160	187	213	240
	17	28^{3}	56^{7}	85	113	142	170	198	227	255
	18	30^{0}	60^{0}	90	120	150	180	210	240	270
	19	31^{7}	63^{3}	95	127	158	190	222	253	285
	20	33^{3}	66^{7}	100	133	167	200	233	267	300
	21	35^{0}	70^{0}	105	140	175	210	245	280	315
	22	36^{7}	73^{3}	110	147	183	220	257	293	330
	23	38^{3}	76^{7}	115	153	192	230	268	307	345
	24	40^{0}	80^{0}	120	160	200	240	280	320	360
	25	41^{7}	83^{3}	125	167	208	250	292	333	375
	26	43^{3}	86^{7}	130	173	217	260	303	347	390
	27	45^{0}	90^{0}	135	180	225	270	315	360	405
	28	46^{7}	93^{3}	140	187	233	280	327	373	420
	29	48^{3}	96^{7}	145	193	242	290	338	387	435
	33	55^{0}	110^{0}	165	220	275	330	385	440	495
	63	105^{0}	210^{0}	315	420	525	630	735	840	945
	93	155^{0}	310^{0}	465	620	775	930	1085	1240	1395
Months.	1	50^{0}	100^{0}	150	200	250	300	350	400	450
	2	100^{0}	200^{0}	300	400	500	600	700	800	900
	3	150^{0}	300^{0}	450	600	750	900	1050	1200	1350
	4	200^{0}	400^{0}	600	800	1000	1200	1400	1600	1800
	5	250^{0}	500^{0}	750	1000	1250	1500	1750	2000	2250
	6	300^{0}	600^{0}	900	1200	1500	1800	2100	2400	2700
	7	350^{0}	700^{0}	1050	1400	1750	2100	2450	2800	3150
	8	400^{0}	800^{0}	1200	1600	2000	2400	2800	3200	3600
	9	450^{0}	900^{0}	1350	1800	2250	2700	3150	3600	4050
	10	500^{0}	1000^{0}	1500	2000	2500	3000	3500	4000	4500
	11	550^{0}	1100^{0}	1650	2200	2750	3300	3850	4400	4950
Years.	1	600^{0}	1200^{0}	1800	2400	3000	3600	4200	4800	5400
	2	1200^{0}	2400^{0}	3600	4800	6000	7200	8400	9600	10800
	3	1800^{0}	3600^{0}	5400	7200	9000	10800	12600	14400	16200
	4	2400^{0}	4800^{0}	7200	9600	12000	14400	16800	19200	21600
	5	3000^{0}	6000^{0}	9000	12000	15000	18000	21000	24000	27000
Comp'd Int.	2	1236^{0}	2472^{0}	3708	4944	6180	7416	8652	9888	11124
	3	1910^{2}	3820^{3}	5730	7641	9551	11461	13371	15281	17191
	4	2624^{8}	5249^{5}	7874	10499	13124	15749	18373	20998	23623
	5	3382^{3}	6764^{5}	10147	13529	16911	20294	23676	27058	30440
	6	4185^{2}	8370^{4}	12556	16741	20926	25111	29296	33482	37667

	Princip'l	$\$100^{0}$ 10 1	$\$200^{0}$ 20 2	$300 30 3	$400 40 4	$500 50 5	$600 60 6	$700 70 7	$800 80 8	$900 90 9
Days	1	1^{9}	3^{9}	6	8	10	12	14	16	18
	2	3^{9}	7^{8}	12	16	19	23	27	31	35
	3	5^{8}	11^{7}	18	23	29	35	41	47	53
	4	7^{8}	15^{6}	23	31	39	47	54	62	70
	5	9^{7}	19^{4}	29	39	49	58	68	78	88
	6	11^{7}	23^{3}	35	47	58	70	82	93	105
	7	13^{6}	27^{2}	41	54	68	82	95	109	123
	8	15^{6}	31^{1}	47	62	78	93	109	124	140
	9	17^{5}	35^{0}	53	70	88	105	123	140	158
	10	19^{4}	38^{9}	58	78	97	117	136	156	175
	11	21^{4}	42^{8}	64	86	107	128	150	171	193
	12	23^{3}	46^{7}	70	93	117	140	163	187	210
	13	25^{3}	50^{6}	76	101	126	152	177	202	228
	14	27^{2}	54^{4}	82	109	136	163	191	218	245
	15	29^{2}	58^{3}	88	117	146	175	204	233	263
	16	31^{1}	62^{2}	93	124	156	187	218	249	280
	17	33^{1}	66^{1}	99	132	165	198	231	264	298
	18	35^{0}	70^{0}	105	140	175	210	245	280	315
	19	36^{9}	73^{9}	111	148	185	222	259	296	333
	20	38^{9}	77^{8}	117	156	194	233	272	311	350
	21	40^{8}	81^{7}	123	163	204	245	286	327	368
	22	42^{8}	85^{6}	128	171	214	257	299	342	385
	23	44^{7}	89^{4}	134	179	224	268	313	358	403
	24	46^{7}	93^{3}	140	187	233	280	327	373	420
	25	48^{6}	97^{2}	146	194	243	292	340	389	438
	26	50^{6}	101^{1}	152	202	253	303	354	404	455
	27	52^{5}	105^{0}	158	210	263	315	368	420	473
	28	54^{4}	108^{9}	163	218	272	327	381	436	490
	29	56^{4}	112^{8}	169	226	282	338	395	451	508
	33	64^{2}	128^{3}	193	257	321	385	449	513	578
	63	122^{5}	245^{0}	368	490	613	735	858	980	1103
	93	180^{8}	361^{7}	543	723	904	1085	1266	1447	1628
Months	1	58^{3}	116^{7}	175	233	292	350	408	467	525
	2	116^{7}	233^{3}	350	467	583	700	817	933	1050
	3	175^{0}	350^{0}	525	700	875	1050	1225	1400	1575
	4	233^{3}	466^{7}	700	933	1167	1400	1633	1867	2100
	5	291^{7}	583^{3}	875	1167	1458	1750	2042	2333	2625
	6	350^{0}	700^{0}	1050	1400	1750	2100	2450	2800	3150
	7	408^{3}	816^{7}	1225	1633	2042	2450	2858	3267	3675
	8	466^{7}	933^{3}	1400	1867	2333	2800	3267	3733	4200
	9	525^{0}	1050^{0}	1575	2100	2625	3150	3675	4200	4725
	10	583^{3}	1166^{7}	1750	2333	2917	3500	4083	4667	5250
	11	641^{7}	1283^{3}	1925	2567	3208	3850	4492	5133	5775
Years	1	700^{0}	1400^{0}	2100	2800	3500	4200	4900	5600	6300
	2	1400^{0}	2800^{0}	4200	5600	7000	8400	9800	11200	12600
	3	2100^{0}	4200^{0}	6300	8400	10500	12600	14700	16800	18900
	4	2800^{0}	5600^{0}	8400	11200	14000	16800	19600	22400	25200
	5	3500^{0}	7000^{0}	10500	14000	17500	21000	24500	28000	31500
Comp'd Int.	2	1449^{0}	2898^{0}	4347	5796	7245	8694	10143	11592	13041
	3	2250^{4}	4500^{9}	6751	9002	11252	13503	15753	18003	20254
	4	3108^{0}	6215^{9}	9324	12432	15540	18648	21756	24864	27972
	5	4025^{6}	8051^{0}	12077	16102	20128	24153	28179	32204	36230
	6	5007^{8}	10014^{6}	15022	20029	25037	30044	35051	40058	45066

Princip'l	$100^0 10 1	$200^0 20 2	$300 30 3	$400 40 4	$500 50 5	$600 60 6	$700 70 7	$800 80 8	$900 90 9
Days.									
1	2^2	4^4	7	9	11	13	16	18	20
2	4^4	8^9	13	18	22	27	31	36	40
3	6^7	13^3	20	27	33	40	47	53	60
4	8^9	17^8	27	36	44	53	62	71	80
5	11^1	22^2	33	44	56	67	78	89	100
6	13^3	26^7	40	53	67	80	93	107	120
7	15^6	31^1	47	62	78	93	109	124	140
8	17^8	35^6	53	71	89	107	124	142	160
9	20^0	40^0	60	80	100	120	140	160	180
10	22^2	44^4	67	89	111	133	156	178	200
11	24^4	48^9	73	98	122	147	171	196	220
12	26^7	53^3	80	107	133	160	187	213	240
13	28^9	57^8	87	116	144	173	202	231	260
14	31^1	62^2	93	124	156	187	218	249	280
15	33^3	66^7	100	133	167	200	233	267	300
16	35^6	71^1	107	142	178	213	249	284	320
17	37^8	75^6	113	151	189	227	264	302	340
18	40^0	80^0	120	160	200	240	280	320	360
19	42^2	84^4	127	169	211	253	296	338	380
20	44^4	88^9	133	178	222	267	311	356	400
21	46^7	93^3	140	187	233	280	327	373	420
22	48^9	97^8	147	196	244	293	342	391	440
23	51^1	102^2	153	204	256	307	358	409	460
24	53^3	106^7	160	213	267	320	373	427	480
25	55^6	111^1	167	222	278	333	389	444	500
26	57^8	115^6	173	231	289	347	404	462	520
27	60^0	120^0	180	240	300	360	420	480	540
28	62^2	124^4	187	249	311	373	436	498	560
29	64^4	128^9	193	258	322	387	451	516	580
33	73^3	146^7	220	293	367	440	513	587	660
63	140^0	280^0	420	560	700	840	980	1120	1260
93	206^7	413^3	620	827	1033	1240	1447	1653	1860
Months.									
1	66^7	133^3	200	267	333	400	467	533	600
2	133^3	266^7	400	533	667	800	933	1067	1200
3	200^0	400^0	600	800	1000	1200	1400	1600	1800
4	266^7	533^3	800	1067	1333	1600	1867	2133	2400
5	333^3	666^7	1000	1333	1667	2000	2333	2667	3000
6	400^0	800^0	1200	1600	2000	2400	2800	3200	3600
7	466^7	933^3	1400	1867	2333	2800	3267	3733	4200
8	533^3	1066^7	1600	2133	2667	3200	3733	4267	4800
9	600^0	1200^0	1800	2400	3000	3600	4200	4800	5400
10	666^7	1333^3	2000	2667	3333	4000	4667	5333	6000
11	733^3	1466^7	2200	2933	3667	4400	5133	5867	6600
Years.									
1	800^0	1600^0	2400	3200	4000	4800	5600	6400	7200
2	1600^0	3200^0	4800	6400	8000	9600	11200	12800	14400
3	2400^0	4800^0	7200	9600	12000	14400	16800	19200	21600
4	3200^0	6400^0	9600	12800	16000	19200	22400	25600	28800
5	4000^0	8000^0	12000	16000	20000	24000	28000	32000	36000
Comp'd Int.									
2	1664^0	3328^0	4992	6656	8320	9984	11648	13312	14976
3	2597^1	5194^2	7791	10388	12986	15583	18180	20777	23374
4	3604^9	7209^8	10815	14420	18024	21629	25234	28839	32444
5	4693^8	9386^6	14080	18773	23466	28160	32853	37546	42240
6	5868^7	11737^5	17606	23475	29344	35212	41081	46950	52819

Principal	$100^{0} / 10 / 1	$200^{0} / 20 / 2	$300 / 30 / 3	$400 / 40 / 4	$500 / 50 / 5	$600 / 60 / 6	$700 / 70 / 7	$800 / 80 / 8	$900 / 90 / 9
Days. 1	2^{8}	5^{6}	8	11	14	17	19	22	25
2	5^{6}	11^{1}	17	22	28	33	39	44	50
3	8^{3}	16^{7}	25	33	42	50	58	67	75
4	11^{1}	22^{2}	33	44	56	67	78	89	100
5	13^{9}	27^{8}	42	56	69	83	97	111	125
6	16^{7}	33^{3}	50	67	83	100	117	133	150
7	19^{4}	38^{9}	58	78	97	117	136	156	175
8	22^{2}	44^{4}	67	89	111	133	156	178	200
9	25^{0}	50^{0}	75	100	125	150	175	200	225
10	27^{8}	55^{6}	83	111	139	167	194	222	250
11	30^{6}	61^{1}	92	122	153	183	214	244	275
12	33^{3}	66^{7}	100	133	167	200	233	267	300
13	36^{1}	72^{2}	108	144	181	217	253	289	325
14	38^{9}	77^{8}	117	156	194	233	272	311	350
15	41^{7}	83^{3}	125	167	208	250	292	333	375
16	44^{4}	88^{9}	133	178	222	267	311	356	400
17	47^{2}	94^{4}	142	189	236	283	331	378	425
18	50^{0}	100^{0}	150	200	250	300	350	400	450
19	52^{8}	105^{6}	158	211	264	317	369	422	475
20	55^{6}	111^{1}	167	222	278	333	389	444	500
21	58^{3}	116^{7}	175	233	292	350	408	467	525
22	61^{1}	122^{2}	183	244	306	367	428	489	550
23	63^{9}	127^{8}	192	256	319	383	447	511	575
24	66^{7}	133^{3}	200	267	333	400	467	533	600
25	69^{4}	138^{9}	208	278	347	417	486	556	625
26	72^{2}	144^{4}	217	289	361	433	506	578	650
27	75^{0}	150^{0}	225	300	375	450	525	600	675
28	77^{8}	155^{6}	233	311	389	467	544	622	700
29	80^{6}	161^{1}	242	322	403	483	564	644	725
33	91^{7}	183^{3}	275	367	458	550	642	733	825
63	175^{0}	350^{0}	525	700	875	1050	1225	1400	1575
93	258^{3}	516^{7}	775	1033	1292	1550	1808	2067	2325
Months. 1	83^{3}	166^{7}	250	333	417	500	583	667	750
2	166^{7}	333^{3}	500	667	833	1000	1167	1333	1500
3	250^{0}	500^{0}	750	1000	1250	1500	1750	2000	2250
4	333^{3}	666^{7}	1000	1333	1667	2000	2333	2667	3000
5	416^{7}	833^{3}	1250	1667	2083	2500	2917	3333	3750
6	500^{0}	1000^{0}	1500	2000	2500	3000	3500	4000	4500
7	583^{3}	1166^{7}	1750	2333	2917	3500	4083	4667	5250
8	666^{7}	1333^{3}	2000	2667	3333	4000	4667	5333	6000
9	750^{0}	1500^{0}	2250	3000	3750	4500	5250	6000	6750
10	833^{3}	1666^{7}	2500	3333	4167	5000	5833	6667	7500
11	916^{7}	1833^{3}	2750	3667	4583	5500	6417	7333	8250
Years. 1	1000^{0}	2000^{0}	3000	4000	5000	6000	7000	8000	9000
2	2000^{0}	4000^{0}	6000	8000	10000	12000	14000	16000	18000
3	3000^{0}	6000^{0}	9000	12000	15000	18000	21000	24000	27000
4	4000^{0}	8000^{0}	12000	16000	20000	24000	28000	32000	36000
5	5000^{0}	10000^{0}	15000	20000	25000	30000	35000	40000	45000
Comp'd Int. 2	2100^{0}	4200^{0}	6300	8400	10500	12600	14700	16800	18900
3	3310^{0}	6620^{0}	9930	13240	16550	19860	23170	26480	29790
4	4641^{0}	9282^{0}	13923	18564	23205	27846	32487	37128	41769
5	6105^{1}	12210^{2}	18315	24420	30526	36631	42736	48841	54946
6	7715^{6}	15431^{2}	23147	30862	38578	46294	54009	61725	69441

TIME TABLE for finding the exact number of Days between two dates; and the date a Note or Bill matures.

Jan.	1	2	3	4	5	6	7	8	9	10	11	12	13	14	15	16	17	18	19	20	21	22	23	24	25	26	27	28	29	30	31
Feb.	32	33	34	35	36	37	38	39	40	41	42	43	44	45	46	47	48	49	50	51	52	53	54	55	56	57	58	59			
Mar.	60	61	62	63	64	65	66	67	68	69	70	71	72	73	74	75	76	77	78	79	80	81	82	83	84	85	86	87	88	89	90
April	91	92	93	94	95	96	97	98	99	100	101	102	103	104	105	106	107	108	109	110	111	112	113	114	115	116	117	118	119	120	
May	121	122	123	124	125	126	127	128	129	130	131	132	133	134	135	136	137	138	139	140	141	142	143	144	145	146	147	148	149	150	151
June	152	153	154	155	156	157	158	159	160	161	162	163	164	165	166	167	168	169	170	171	172	173	174	175	176	177	178	179	180	181	
July	182	183	184	185	186	187	188	189	190	191	192	193	194	195	196	197	198	199	200	201	202	203	204	205	206	207	208	209	210	211	212
Aug.	213	214	215	216	217	218	219	220	221	222	223	224	225	226	227	228	229	230	231	232	233	234	235	236	237	238	239	240	241	242	243
Sep.	244	245	246	247	248	249	250	251	252	253	254	255	256	257	258	259	260	261	262	263	264	265	266	267	268	269	270	271	272	273	
Oct.	274	275	276	277	278	279	280	281	282	283	284	285	286	287	288	289	290	291	292	293	294	295	296	297	298	299	300	301	302	303	304
Nov.	305	306	307	308	309	310	311	312	313	314	315	316	317	318	319	320	321	322	323	324	325	326	327	328	329	330	331	332	333	334	
Dec.	335	336	337	338	339	340	341	342	343	344	345	346	347	348	349	350	351	352	353	354	355	356	357	358	359	360	361	362	363	364	365

	1	2	3	4	5	6	7	8	9	10	11	12	13	14	15	16	17	18	19	20	21	22	23	24	25	26	27	28	29	30	31
Jan.	366	367	368	369	370	371	372	373	374	375	376	377	378	379	380	381	382	383	384	385	386	387	388	389	390	391	392	393	394	395	396
Feb.	397	398	399	400	401	402	403	404	405	406	407	408	409	410	411	412	413	414	415	416	417	418	419	420	421	422	423	424			
Mar.	425	426	427	428	429	430	431	432	433	434	435	436	437	438	439	440	441	442	443	444	445	446	447	448	449	450	451	452	453	454	455
April	456	457	458	459	460	461	462	463	464	465	466	467	468	469	470	471	472	473	474	475	476	477	478	479	480	481	482	483	484	485	
May	486	487	488	489	490	491	492	493	494	495	496	497	498	499	500	501	502	503	504	505	506	507	508	509	510	511	512	513	514	515	516
June	517	518	519	520	521	522	523	524	525	526	527	528	529	530	531	532	533	534	535	536	537	538	539	540	541	542	543	544	545	546	
July	547	548	549	550	551	552	553	554	555	556	557	558	559	560	561	562	563	564	565	566	567	568	569	570	571	572	573	574	575	576	577
Aug.	578	579	580	581	582	583	584	585	586	587	588	589	590	591	592	593	594	595	596	597	598	599	600	601	602	603	604	605	606	607	608
Sep.	609	610	611	612	613	614	615	616	617	618	619	620	621	622	623	624	625	626	627	628	629	630	631	632	633	634	635	636	637	638	
Oct.	639	640	641	642	643	644	645	646	647	648	649	650	651	652	653	654	655	656	657	658	659	660	661	662	663	664	665	666	667	668	669
Nov.	670	671	672	673	674	675	676	677	678	679	680	681	682	683	684	685	686	687	688	689	690	691	692	693	694	695	696	697	698	699	
Dec.	700	701	702	703	704	705	706	707	708	709	710	711	712	713	714	715	716	717	718	719	720	721	722	723	724	725	726	727	728	729	730

LUMBER TABLE, showing the contents of BOARDS, SCANTLINGS, JOISTS, SILLS, etc.—in square ft. & in

Board Measure.

Width. / Length.	12	14	16	18	20
4 in.	4	4^{8}	5^{4}	6	6^{8}
5 "	5	5^{10}	6^{8}	7^{6}	8^{4}
6 "	6	7	8	9	10
7 "	7	8^{2}	9^{4}	10^{6}	11^{8}
8 "	8	9^{4}	10^{8}	12	13^{4}
9 "	9	10^{6}	12	13^{6}	15
10 "	10	11^{8}	13^{4}	15	16^{8}
11 "	11	12^{10}	14^{8}	16^{6}	18^{4}
12 "	12	14	16	18	20
13 "	13	15^{2}	17^{4}	19^{6}	21^{8}
14 "	14	16^{4}	18^{8}	21	23^{4}
15 "	15	17^{6}	20	22^{6}	25
16 "	16	18^{8}	21^{4}	24	26^{8}
17 "	17	19^{10}	22^{8}	25^{6}	28^{4}
18 "	18	21	24	27	30
19 "	19	22^{2}	25^{4}	28^{6}	31^{8}
20 "	20	23^{4}	26^{8}	30	33^{4}
21 "	21	24^{6}	28	31^{6}	35
22 "	22	25^{8}	29^{4}	33	36^{8}
23 "	23	26^{10}	30^{8}	34^{6}	38^{4}
24 "	24	28	32	36	40
25 "	25	29^{2}	33^{4}	37^{6}	41^{8}
26 "	26	30^{4}	34^{8}	39	43^{4}
27 "	27	31^{6}	36	40^{6}	45
28 "	28	32^{8}	37^{4}	42	46^{8}
29 "	29	33^{10}	38^{8}	43^{6}	48^{4}
30 "	30	35	40	45	50

Joist, Sill and Scantling Measure.

Size in. / Length.	12	14	16	18	20	22	24	26	28	30
2 by 2	4	4^{8}	5^{4}	6	6^{8}	7^{4}	8	8^{8}	9^{4}	10
2 " 3	6	7	8	9	10	11	12	13	14	15
2 " 4	8	9^{4}	10^{8}	12	13^{4}	14^{8}	16	17^{4}	18^{8}	20
2 " 5	10	11^{8}	13^{4}	15	16^{8}	18^{4}	20	21^{8}	23^{4}	25
2 " 6	12	14	16	18	20	22	24	26	28	30
2 " 8	16	18^{8}	21^{4}	24	26^{8}	29^{4}	32	34^{8}	37^{4}	40
2 " 10	20	23^{4}	26^{8}	30-	33^{4}	36^{8}	40	43^{4}	46^{8}	50
2 " 12	24	28	32	36	40	44	48	52	56	60
3 " 3	9	10^{6}	12	13^{6}	15	16^{6}	18	19^{6}	21	22^{6}
3 " 4	12	14	16	18	20	22	24	26	28	30
3 " 6	18	21	24	27	30	33	36	39	42	45
3 " 8	24	28	32	36	40	44	48	52	56	60
3 " 10	30	35	40	45	50	55	60	65	70	75
3 " 12	36	42	48	54	60	66	72	78	84	90
4 " 4	16	18^{8}	21^{4}	24	26^{8}	29^{4}	32	34^{8}	37^{4}	40
4 " 5	20	23^{4}	26^{8}	30	33^{4}	36^{8}	40	43^{4}	46^{8}	50
4 " 6	24	28	32	36	40	44	48	52	56	60
4 " 8	32	37^{4}	42^{8}	48	53^{4}	58^{8}	64	69^{4}	74^{8}	80
4 " 10	40	46^{8}	53^{4}	60	66^{8}	73^{4}	80	86^{8}	93^{4}	100
4 " 12	48	56	64	72	80	88	96	104	112	120
6 " 6	36	42	48	54	60	66	72	78	84	90
6 " 8	48	56	64	72	80	88	96	104	112	120
8 " 8	64	74^{8}	85^{4}	96	106^{8}	117^{4}	128	138^{8}	149^{4}	160
8 " 10	80	93^{4}	106^{8}	120	133^{4}	146^{8}	160	173^{4}	186^{8}	200
10 " 10	100	116^{8}	133^{4}	150	166^{8}	183^{4}	200	216^{8}	233^{4}	250
10 " 12	120	140	160	180	200	220	240	260	280	300
12 " 12	144	168	192	216	240	264	288	312	336	360

Table showing contents of Saw Logs in Inch board measure.

L'gth. / Diameter in Inches.	12	14	16	18	20	22	24	26	28	30
12	50	58	67	75	83	92	100	108	117	125
13	63	73	84	95	105	116	126	137	147	158
14	77	90	103	116	129	142	155	167	180	193
15	93	109	124	140	155	171	186	202	217	233
16	110	128	147	165	183	202	220	238	257	275
17	128	150	171	193	214	236	257	279	300	321
18	148	173	197	222	247	271	296	321	345	370
19	169	198	226	254	283	311	339	367	395	423
20	192	224	256	288	320	352	384	416	448	480
21	216	253	289	325	361	397	433	469	505	541
22	242	282	323	363	403	444	484	524	565	605
23	269	314	359	404	449	494	539	583	628	673
24	298	348	397	447	497	546	596	646	695	745
25	328	383	438	493	547	602	657	711	766	821
26	360	420	480	540	600	660	720	780	840	900
27	393	458	524	590	656	721	787	852	918	983
28	428	500	571	642	714	785	856	928	999	1070
29	464	542	620	697	775	852	929	1007	1084	1161
30	502	586	669	753	837	920	1004	1088	1171	1255
31	541	632	722	812	903	993	1083	1173	1263	1353
32	582	679	776	873	970	1067	1164	1261	1358	1455
33	624	729	833	937	1041	1145	1249	1353	1457	1561
34	668	779	890	1002	1113	1224	1336	1447	1558	1670
35	713	832	951	1070	1189	1308	1427	1545	1664	1783
36	760	887	1013	1140	1267	1393	1520	1647	1773	1900
37	808	943	1078	1213	1347	1482	1617	1751	1886	2021
38	858	1001	1144	1287	1430	1573	1716	1859	2002	2145
39	909	1060	1212	1364	1515	1667	1819	1970	2122	2273
40	962	1122	1282	1443	1603	1763	1924	2084	2244	2405
41	1016	1186	1355	1525	1694	1864	2033	2202	2372	2541
42	1072	1250	1429	1608	1786	1965	2144	2322	2501	2680

Table showing contents of Cisterns, Tanks, &c. in bbl. of 31½ gls.

Depth. / Diameter in Ft.	5	6	7	8	9	10	11	12	13	14	15	16	18	20
4	15	18	21	24	27	30	33	36	39	42	45	48	54	60
4½	19	23	26	30	34	38	42	45	49	53	57	60	68	76
5	23	28	32	37	42	47	51	56	61	65	70	75	84	93
5½	28	34	39	45	51	56	62	68	73	79	85	90	101	113
6	34	40	47	54	60	67	74	81	87	94	101	107	121	134
6½	39	47	55	63	71	79	87	95	102	110	118	126	142	158
7	46	55	64	73	82	91	101	110	119	128	137	146	165	183
7½	52	63	73	84	94	105	115	126	136	147	157	168	189	210
8	60	72	84	95	107	119	131	143	155	167	179	191	215	239
8½	67	81	94	108	121	135	148	162	175	189	202	216	243	269
9	76	91	106	121	136	151	166	181	196	211	227	242	272	302
9½	84	101	118	135	151	168	185	202	219	236	252	269	303	337
10	93	112	131	149	168	187	205	224	242	261	280	298	336	373
11	113	135	158	181	203	226	248	271	293	316	339	361	406	451
12	134	161	188	215	242	269	295	322	349	376	403	430	483	537
13	158	189	221	252	284	315	347	378	410	441	473	504	567	630
14	183	219	256	292	329	366	402	439	475	512	548	585	658	731
15	210	252	294	336	378	420	462	504	546	587	629	671	755	839
16	239	286	334	382	430	477	525	573	621	668	716	764	859	955
18	302	363	423	483	544	604	665	725	786	846	906	967	1088	1209
20	373	448	522	597	671	746	821	895	970	1044	1119	1194	1343	1492

L'gth. / Width in Feet.	8	9	10	11	12	13	14	15	16	18	20	21
3	192	217	241	265	289	313	338	362	386	434	482	506
3½	225	253	281	309	338	366	394	422	450	506	563	591
4	257	289	321	354	386	418	450	482	514	579	643	675
4½	289	325	362	398	434	470	506	542	579	651	723	759
5	321	362	402	442	482	522	563	603	643	723	804	844
5½	354	398	442	486	530	575	619	663	707	796	884	928
6	386	434	482	530	579	627	675	723	771	868	964	1013
6½	418	470	522	575	627	679	731	783	836	940	1045	1097
7	450	506	563	619	675	731	788	844	900	1013	1125	1181
7½	482	542	603	663	723	783	844	904	964	1085	1205	1266
8	514	579	643	707	771	836	900	964	1029	1157	1286	1350
8½	546	615	683	751	820	888	956	1025	1093	1229	1366	1434
9	579	651	723	796	868	940	1013	1085	1157	1302	1446	1519
9½	611	687	763	840	916	992	1069	1145	1221	1374	1527	1603
10	643	723	804	884	964	1045	1125	1205	1286	1446	1607	1688
11	707	796	884	972	1061	1149	1238	1326	1414	1591	1768	1856
12	771	868	964	1061	1157	1254	1350	1446	1543	1736	1929	2025

Table showing contents of Corn-cribs 10 ft. high—Corn in ear.*

L'gth. / Width in Feet.	10	11	12	14	16	18	20	22	24	26	28	30	32
3	135	149	162	189	216	243	270	297	324	351	378	405	432
3½	158	173	189	221	252	284	315	347	378	410	441	473	504
4	180	198	216	252	288	324	360	396	432	468	504	540	576
4½	203	223	243	284	324	365	405	446	486	527	567	608	648
5	225	248	270	315	360	405	450	495	540	585	630	675	720
5½	248	272	297	347	396	446	495	545	594	644	693	743	792
6	270	297	324	378	432	486	540	594	648	702	756	810	864
6½	293	322	351	410	468	527	585	644	702	761	819	878	936
7	315	347	378	441	504	567	630	693	756	819	882	945	1008
7½	338	371	405	473	540	608	675	743	810	878	945	1013	1080
8	360	396	432	504	576	648	720	792	864	936	1008	1080	1152
8½	383	421	459	536	612	689	765	842	918	995	1071	1148	1224
9	405	446	486	567	648	729	810	891	972	1053	1134	1215	1296
9½	428	470	513	599	684	770	855	941	1026	1112	1197	1283	1368
10	450	495	540	630	720	810	900	990	1080	1170	1260	1350	1440
11	495	545	594	693	792	891	990	1089	1188	1287	1386	1485	1584
12	540	594	648	756	864	972	1080	1188	1296	1404	1512	1620	1728

The top lines indicate the length, the left hand columns the width. A bin 7 ft. wide and 16 ft. long will hold 900 bu. of grain or 504 bu. of corn in the ear, supposing it to be *ten* ft. high. When *more* or *less* than 10 ft. high, *cut off the right hand figure and multiply by the given height.* For instances, a corn crib 8 ft. wide, 18 ft. long and *nine* ft. high, contains 583 bu.

$$\begin{array}{r} 64.8 \\ 9 \\ \hline 583.2 \end{array}$$

A Wagon-bed, 3 ft. wide, 10 ft. long and 15 inches deep, will hold 30 bu. and 1 tenth. Cutting off the right hand figure from the number corresponding to the width and length, gives the contents of a body 12 inches deep. Then *add* to this number such part of itself as the depth *over* 12 inches, is part of 12. Thus, for 15 in. add ¼; for 16 inches ⅓; for 18 in. ½, etc. Or *multiply* the number found in table *by the depth in inches, divide by* 12 *and cut off right hand figure.*

$$\begin{array}{r} 4)24.1 \\ 6.0 \\ \hline 30.1 \end{array}$$

*Rules for measuring corn in the ear, vary all the way from 3456 to 4320 cubic inches to the bushel. No rule can be laid down that will tally in all kinds of corn. The above table, and rule on page 67 are based on 3840 cubic inches to the bu., which is considered as reliable as a *general* rule can possibly be, and will hold out when corn is sound.

WAGES Table for Days & Hours at given rates per Week.

	Rate	$3	3½	$4	4½	$5	5½	$6	6½	$7	7½	$8	$9	10	11	12
Hours.	1	5	6	7	8	8	9	.10	.11	.12	.13	.13	.15	.17	.18	.20
	2	.10	.12	.13	.15	.17	.18	.20	.22	.23	.25	.27	.30	.33	.37	.40
	3	.15	.18	.20	.23	.25	.28	.30	.33	.35	.38	.40	.45	.50	.55	.60
	4	.20	.23	.27	.30	.33	.37	.40	.43	.47	.50	.53	.60	.67	.73	.80
	5	.25	.29	.33	.38	.42	.46	.50	.54	.58	.63	.67	.75	.83	.92	1.00
	6	.30	.35	.40	.45	.50	.55	.60	.65	.70	.75	.80	.90	1.00	1.10	1.20
	7	.35	.41	.47	.53	.58	.64	.70	.76	.82	.88	.93	1.05	1.17	1.28	1.40
	8	.40	.47	.53	.60	.67	.73	.80	.87	.93	1.00	1.07	1.20	1.33	1.47	1.60
	9	.45	.53	.60	.68	.75	.83	.90	.98	1.05	1.13	1.20	1.35	1.50	1.65	1.80
Days.	1	.50	.58	.67	.75	.83	.92	1.00	1.08	1.17	1.25	1.33	1.50	1.67	1.83	2.00
	2	1.00	1.17	1.33	1.50	1.67	1.83	2.00	2.17	2.33	2.50	2.67	3.00	3.33	3.67	4.00
	3	1.50	1.75	2.00	2.25	2.50	2.75	3.00	3.25	3.50	3.75	4.00	4.50	5.00	5.50	6.00
	4	2.00	2.33	2.67	3.00	3.33	3.67	4.00	4.33	4.67	5.00	5.33	6.00	6.67	7.33	8.00
	5	2.50	2.92	3.33	3.75	4.17	4.58	5.00	5.42	5.83	6.25	6.67	7.50	8.33	9.17	$10.

Table showing the WAGES for Days at given rates per Month.

	Rate.	$14.	$15.	$16.	$17.	$18.	$19.	$20.	$21.	$22.	$23.	$24.	$25.
Days.	1	.54	.58	.62	.65	.69	.73	.77	.81	.85	.88	.92	.96
	2	1.08	1.15	1.23	1.31	1.38	1.46	1.54	1.62	1.69	1.77	1.85	1.92
	3	1.62	1.73	1.85	1.96	2.08	2.19	2.31	2.42	2.54	2.65	2.77	2.88
	4	2.15	2.31	2.46	2.62	2.77	2.92	3.08	3.23	3.38	3.54	3.69	3.85
	5	2.69	2.88	3.08	3.27	3.46	3.65	3.85	4.04	4.23	4.42	4.62	4.81
	6	3.23	3.46	3.69	3.92	4.15	4.38	4.62	4.85	5.08	5.31	5.54	5.77
	7	3.77	4.04	4.31	4.58	4.85	5.12	5.38	5.65	5.92	6.19	6.46	6.73
	8	4.31	4.62	4.92	5.23	5.54	5.85	6.15	6.46	6.77	7.08	7.38	7.69
	9	4.85	5.19	5.54	5.88	6.23	6.58	6.92	7.27	7.62	7.96	8.31	8.65
	10	5.38	5.77	6.15	6.54	6.92	7.31	7.69	8.08	8.46	8.85	9.23	9.62
	11	5.92	6.35	6.77	7.19	7.62	8.04	8.46	8.88	9.31	9.73	10.15	10.58
	12	6.46	6.92	7.38	7.85	8.31	8.77	9.23	9.69	10.15	10.62	11.08	11.54
	13	7.00	7.50	8.00	8.50	9.00	9.50	10.00	10.50	11.00	11.50	12.00	12.50
	14	7.54	8.08	8.62	9.15	9.69	10.23	10.77	11.31	11.85	12.38	12.92	13.46
	15	8.08	8.65	9.23	9.81	10.38	10.96	11.54	12.12	12.69	13.27	13.85	14.42
	16	8.62	9.23	9.85	10.46	11.08	11.69	12.31	12.92	13.54	14.15	14.77	15.38
	17	9.15	9.81	10.46	11.12	11.77	12.42	13.08	13.73	14.38	15.04	15.69	16.35
	18	9.69	10.38	11.08	11.77	12.46	13.15	13.85	14.54	15.23	15.92	16.62	17.31
	19	10.23	10.96	11.69	12.42	13.15	13.88	14.62	15.35	16.08	16.81	17.54	18.27
	20	10.77	11.54	12.31	13.08	13.85	14.62	15.38	16.15	16.92	17.69	18.46	19.23
	21	11.31	12.12	12.92	13.73	14.54	15.35	16.15	16.96	17.77	18.58	19.38	20.19
	22	11.85	12.69	13.54	14.38	15.23	16.08	16.92	17.77	18.62	19.46	20.31	21.15
	23	12.38	13.27	14.15	15.04	15.92	16.81	17.69	18.58	19.46	20.35	21.23	22.12
	24	12.92	13.85	14.77	15.69	16.62	17.54	18.46	19.38	20.31	21.23	22.15	23.08
	25	13.46	14.42	15.38	16.35	17.31	18.27	19.23	20.19	21.15	22.12	23.08	24.04
	26	14.00	15.00	16.00	17.00	18.00	19.00	20.00	21.00	22.00	23.00	24.00	25.00

Table showing the equivalent DECIMALS of Common Fractions.

Com. Frac.	1/2	1/3	2/3	1/4	3/4	1/5	2/5	3/5	4/5
Deci. "	.5	$.33^{33}$	$.66^{66}$	.25	.75	.2	.4	.6	.8
Com. Frac.	1/6	5/6	1/8	3/8	5/8	7/8	1/12	5/12	7/12
Deci. "	$.16^{66}$	$.83^{33}$	$.12^{5}$	$.37^{5}$	$.62^{5}$	$.87^{5}$	$.08^{33}$	$.41^{66}$	$.58^{33}$
Com. Frac.	11/12	1/16	3/16	5/16	7/16	9/16	11/16	13/16	15/16
Deci. "	$.91^{66}$	$.06^{25}$	$.18^{75}$	$.31^{25}$	$.43^{75}$	$.56^{25}$	$.68^{75}$	$.81^{25}$	$.93^{75}$

ADDITION.

Addition is the process of finding the sum of two or more numbers.

Addition of Decimals. RULE.—*Write the numbers so that the decimal points shall stand directly under each other. Add as in simple addition, and place the decimal point in the sum, directly under the points above.*

Add 9.5, 56.25, 672.875, and 3008.3125.

```
        9.5
       56.25
      672.875
     3008.3125
Ans. 3746.9375 Sum, or
               Amount.
```

All who would become proficient in adding long columns of figures, should practice the following method.

Begin at the foot of the right hand column and add, naming results *only;* thus, 15, 20, 29, 35, 42, 50, 54; set down the 4, add the 5 (tens) to the next column and proceed in the same manner, 14, 17, 23, 31, 36, 40, 49, 56. This is much more philosophic, and considerably quicker, than to crawl up a column in the following manner; thus, 7 and 8 are 15, and 15 and 5 are 20, and 20 and 9 are 29, and so on.

```
 74
 98
 47
 56
 89
 65
 38
 97
564
```

Always add the carrying figure to the next column *on commencing*, and when the columns are long, it is well to set it down, as it will often save the trouble of going over the work already performed.

To test addition: *Add the columns in opposite directions.*

SUBTRACTION.

Subtraction is the process of finding the difference between two numbers.

Subtraction of Decimals. RULE.—*Write the numbers so that the decimal points shall stand directly under each other. Subtract as in whole numbers, and place the decimal point in the remainder, directly under the points above.*

From 843.75 take 597.625.		843.750	*Minuend.*
		597.625	*Subtrahend.*
	Ans.	246.125	*Difference, or Remainder.*

Two or more numbers may be taken from another, at a single operation, *by writing in the remainder, such figures, as added to the sum of each column of the subtrahends, will make its right hand figure equal to the corresponding figure in the minuend.*

A man who had an annual income of $2367, paid for board $645, for clothing $463, and $386 for incidental expenses. How much money had he left at the end of the year?		2367	*Min.*
		645	*Sub.*
		463	"
		386	"
	Ans.	$873	*Rem.*

Write the subtrahends under the minuend and proceed, thus, saying 6, 3 and 5 are 14, and 3 (written in the remainder) are 17, a number whose right hand figure is equal to the corresponding figure in the minuend. 1 to carry to 8, 6 and 4 are 19, and 7 (written in the rem.) are 26. 2 to carry to 3, 4 and 6 are 15, and 8 (written in the rem.) are 23.

In balancing accounts this method may be applied with practical advantage, in finding the difference between the two sides.

Add the larger side in the usual manner; then begin at the top of the smaller side, *adding downward, and writing such figures in the base, as are needed to produce the required balance.*

Dr. WILLIAM WEAVER. Cr.

1873.				1873.			
Jan. 7,	To Mdse.,	$34	75	Jan. 25,	By Cash,	$56	25
" 25,	" "	41	67	Mar. 15,	" "	23	75
Mar. 4,	" "	17	45	" 30,	" Balance,	$13	87
		$93	87			$93	87

In this account we first add the larger (dr.) side, then begin at the top of the other side, adding downward thus, 5+5=10, and 7 (the figure required to produce the balance) are 17; write down the 7 and carry the 1. 1+2+7=10, and 8 (written in the balance) are 18. 1+6+3=10, and 3 (written in the balance) are 13. 1+5+2=8, and 1 (written in the balance are 9.)

To test subtraction: *Add the difference and subtrahend together—the sum must equal the minuend.*

MULTIPLICATION.

Multiplication is the process of taking one number as many times as there are units in another.

Multiplication of Decimals. RULE.—*Multiply as in whole numbers, and point off as many decimal places in the product as there are decimal places in both multiplicand and multiplier. If there be not so many figures in the product, supply the deficiency by prefixing ciphers.*

Multiply 46.75 by 20.5.

```
       46.75  Multiplicand.
        20.5  Multiplier.
      ------
      23375
     9350
     -------
Ans. 958.375  Product.
```

Multiply .25 by .25. $.25 \times .25 = .0625$ Ans.

To multiply by 10, 100, 1000, etc.: *Annex as many ciphers to the multiplicand as there are ciphers in the multiplier.*

If the multiplicand is a decimal number, *remove the decimal point as many places to the right as there are ciphers in the multiplier, annexing ciphers if necessary.*

Multiply 435 by 100. $435 \times 100 = 43500$ Ans.

Multiply 6.25 by 1000. $6.25 \times 1000 = 6250$ Ans.

When there are ciphers on the right of the multiplicand and multiplier, *multiply the significant figures together and annex as many ciphers to the product as there are ciphers on the right of both factors.*

Multiply 6700 by 480.

```
        6700
         480
        ----
        536
       268
     -------
Ans. 3216000
```

To multiply a whole number by a fraction: *Multiply the*

whole number by the numerator of the fraction, and divide the product by the denominator.

Multiply 56 by $\frac{3}{4}$.

```
          56
           3
      4) 168
Ans.      42
```

To test multiplication: *Divide the product by the multiplier—the quotient must equal the multiplicand.*

DIVISION.

Division is the process of finding how many times one number is equal to another.

Division of Decimals. RULE.—*Divide as in whole numbers, annexing ciphers to the dividend if necessary, and point off as many decimal places from the quotient as the decimal places in the dividend exceed those in the divisor. If there be not so many places in the quotient, supply the deficiency by prefixing ciphers.*

Divide 93.5 by 6.75.

See "Short Method of Division," page 37.

```
Divisor. Dividend. Quotient.
6.75) 93.5000 (13.85+ Ans.
       26 00
        5 750
          3500
           125 Remainder.
```

Divide .784 by 24.5.

```
24.5) .7840 (.032 Ans.
       490
```

To divide by 10, 100, 1000, etc.: *Cut off as many figures from the right of the dividend as there are ciphers in the divisor. The figures thus cut off will be decimals.*

If the dividend is a decimal number, *remove the decimal point as many places to the left as there are ciphers in the divisor, prefixing ciphers if necessary.*

Divide 6475 by 10. $6475 \div 10 = 647.5$ Ans.

Divide 8.75 by 100. $8.75 \div 100 = .0875$ Ans.

When there are ciphers on the right of the divisor, *cut off the ciphers on the right of the divisor, and as many places from the right of the integral part of the dividend; divide by the significant part of the divisor, and when the first rejected figure in the dividend is reached, place a decimal point in the quotient, and continue the division as far as required.*

Divide 69375 by 1500.

```
15|00) 693|75 (46.25 Ans.
         93
          37
           75
```

To divide a whole number by a fraction: *Multiply the whole number by the denominator of the fraction, and divide the product by the numerator.*

Divide 64 by $\frac{2}{3}$.

```
          64
           3
      2) 192
Ans.      96
```

To test division: *Multiply the divisor and quotient together, and to the product add the remainder—the result must equal the dividend.*

THE DECIMAL SCALE.

All those who are not already familiar with the principles of numeration, should carefully study the following

TABLE.

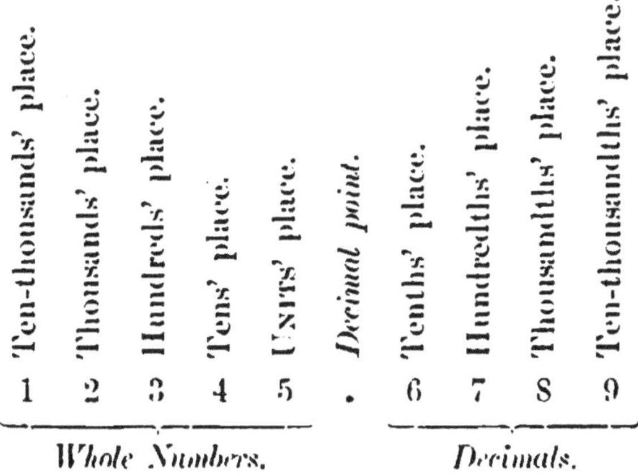

UNITED STATES MONEY.

$1000 place, or order.	$100 place, or order.	$10 place, or order.	DOLLARS' place, or order.	*Decimal point.*	Dimes' place, or order.	Cents' place, or order.	Mills' place, or order.
8	0	6	4	.	3	7	5
Integers.					*Decimals.*		

The ***Decimal point*** is the sign of demarkation between whole numbers or Integers, and decimal fractions.

The *first* place on the *left* of the point, or the *right hand* place in whole numbers, is *units;* the *second* place, *tens;* the *third* place, *hundreds*, etc. The *first* place on the *right* of the point is *tenths;* the *second* place, *hundredths*, and so on.

The United States money system is based on the decimal scale, the *dollars* occupying the *units'* place; the *dimes*, the *tenths'* place; the *cents*, the *hundredths'* place; and the *mills*, the *thousandths'* place.

This and the principles of decimals should be well understood, and committed to memory, by all who would become scientific and proficient calculators.

NOTE.—Be careful to distinguish between *tens* and *tenths*, *hundreds* and *hundredths*, etc., as there is a great difference in the meaning of the two terms.

SHORT METHOD OF MULTIPLICATION.

United States money being based on the decimal system, decimals are involved in nearly all commercial calculations. By the ordinary methods of computing business transactions, a vast amount of decimal figures are usually involved, which are neither essential nor add any thing whatever to the correctness of the required result, since, in practice, orders lower than cents or hundredths are generally disregarded.

By rejecting those decimals in both multiplicand and multiplier which give rise to denominations lower than those required in the answer, the process of multiplication can be greatly shortened, and much useless labor and tedious figuring avoided. This is effected by applying the following philosophic and strictly scientific

RULE.—*Reverse one of the terms and write it for the multiplier, so that its cents' order, or hundredths' place, will fall under the units' place of the multiplicand.*

Reject those figures in the multiplicand which extend to the right of the figure then IN USE *in the multiplier.*

Carry the tens, however, from the product of the nearest rejected figure (multiplied mentally), *and* ONE MORE *when the unit figure of said product is* FIVE *or* OVER.

Set the right hand figures of the partial products in the same column, add, and point off two places from the right of the product—the result will be the answer in units and hundredths, or dollars and cents.

NOTES.—1. It is immaterial which term is taken for the multiplier; usually, however, the cost or price is the most convenient.

2. When the given price is by the 100 or 1000, ascertain its value per unit (mentally): thus, $3 per 100 is 3 cts. per unit; $4 per 1000 is 4 mills per unit etc.: for, according to the rule, the cents' order must fall under the units' place, or *vice versa*, at the price per UNIT.

We will now attempt to elucidate this method of multiplication more fully, and illustrate its practical application. We will, for example, find the value of a lot of steers, weighing 9835 lbs., at $3.18¾ per cwt.

EXPLANATION.—We write the weight 9835, for the multiplicand, and the price per cwt., $3.18¾, with the order of its figures *reversed*, and the fraction ¾, written decimally (.75), for the multiplier. $3 per 100 is 3 cts. per unit; hence, the 3 occupying the *cents'* order must fall under the *units* (5) of the multiplicand, and the other figures, in *reversed* order, to the left of it; mills under tens, etc.

```
      9|8 3 5
    5 7|8 1 3
    ---------
    2 9 5 0 5
        9 8 4
        7 8 6
          6 9
            5
Ans. $3 1 3.4 9
```

We first multiply by the 3 as in ordinary multiplication, obtaining 29505 (cts.) for the first partial product; then mark off the 3 and the 5 above it, by a vertical line, and call them *rejected*. We now proceed to multiply by the 1, beginning with the 3 above it; adding the *tens*, however, from the rejected figure 5, saying, Once 5 is 5, which is equal to ½ of the next higher order—the lowest order retained in the answer. Now, in this system of calculation, ½ or over is counted a *whole* one, and what is *under* is disregarded; thus the gain and loss will be equalized, or nearly so. For this reason we carry from the product of the nearest rejected figure, *one*, when it is 5 or over; *two*, from 15 and over; *three*, from 25 and over, etc. Hence, we say once 3 is 3, and 1 (from the rejected fig. 5) makes 4, which we set under the right hand figure (5) of the first partial product; multiplying on in the usual manner we obtain 984 for the second partial product. We now mark off the 1 and 3, and proceed to multiply by the 8, saying, 8 times 3 (the nearest rejected fig.) are 24, which gives 2 to carry. (The units' fig., 4, being less than ½ of the next higher order, is disregarded.) Hence we proceed: 8 times 8 are 64, and 2 (tens) are 66; multiplying on, we obtain 786 for the third partial product. We next mark off the two 8's, and multiply by the 7, saying, 7 times 8 (the nearest rejected fig.) are 56, which gives 6 to carry. (The units' fig., 6, being over ½, is counted 1.) Thus, 7 times 9 are 63, and 6 (tens) make 69 for the fourth partial product.

Arriving at the last figure in the multiplier, we find no figure over it in the multiplicand; we therefore merely obtain the tens from the nearest rejected figure, saying, 5 times 9 are 45, which gives 5 to carry. We set it in the right hand column of the partial products; in the following examples it is usually added to the product of the preceding figure.

Adding up the partial products, and pointing off two decimal places, the result is $313.49.

The superiority of the short over the ordinary method of multiplication will be more clearly illustrated by the following example. We see that by the short method, all decimals lower than those required in the product are avoided, and yet the answer obtained is sufficiently exact for all practical purposes.

Multiply 8.4125 by 7.6875, retaining only two decimal places in the product.

Ordinary Method.

```
           8.4125
           7.6875
          -------
         |420625
        5|88875
       67|3000
      504|750
     5888|75
     ----|------
Ans. 64.67|109375
```

* See notes on next page.

Short Method.

```
      8.4125
     5786.7
     ------
      5889*
       505
        67
         6
     ------
Ans.  64.67
```

We write the smaller number, in *reversed* order, for the multiplier, so that its units (7) will fall under the 2d decimal place of the multiplicand. If 3 decimals were required in the product, we would write the units under the 3d decimal place; if 4, under the 4th, etc.

NOTES.—1. Occasionally the answer falls short by *one*. This deficiency, however, is obviated by carrying the tens from the second rejected figure to the product of the first, or nearest rejected, when it happens to be nearly 5, 15, 25, 35, etc.

2. In the preceding example we say 7 times 2 (the nearest rejected fig.) are 14, which would give 1 to carry, but by adding the tens from 5 (the 2d rejected fig.) to the 14, it makes 18, and consequently gives *two* to carry.

SHORT METHOD OF DIVISION.

It will be a great advantage to the intelligent student to make himself familiar with the following scientific and practical method of division. It is simple and easy, and does away with about half the figures required by the long method, and in combination with the short method of multiplication, avoids an immense amount of useless and tedious figuring and labor, which is indispensable in the ordinary methods of calculation.

RULE.—*Obtain the first figure in the quotient in the ordinary manner.*

Multiply the first figure of the divisor by this quotient figure, and write such a figure in the remainder as, added to this product, will give an amount whose unit figure is the same as the right hand figure of the partial dividend.

Carry the tens' figure of the amount to the product of the next figure of the divisor, and proceed as before till the entire remainder is obtained.

To this remainder bring down the next figure of the dividend, obtain the second quotient figure and the next remainder in the same manner, and thus proceed till the operation is completed.

EXAMPLES.—Find the average weight of 23 head of hogs weighing 5951 lbs.

```
23) 5951 (258 lbs. Ans.
     135
      201
       17 Remainder.
```

EXPLANATION.—The first figure of the quotient being 2, we multiply the divisor by it, but instead of setting down the product (46) and subtracting it from the partial dividend (59), we simply write down (for the remainder) such figures as are *wanting*, to make the figures of the product *equal* to the corresponding figures of the partial dividend.

Thus, we say, 2 times 3 are 6, and *three*—which is wanting to make 9, the corresponding figure in the partial dividend—we write in the remainder; 2 times 2 are 4, and *one* (written in the rem.) makes 5. To the whole remainder, 13, we bring down the next figure in the dividend (5), making 135. We then proceed: 23 in 135 is contained 5 times; 5 times 3 are 15; here we write a 0 in the remainder, since the unit figure of the product and the right hand figure of the partial dividend are *equal;* 5 times 2 are 10, and 1 (ten) from 15, are 11, and *two* (written in the rem.) are 13. To the remainder, 20, we annex the 1, making 201. 23 in 201, 8 times; 8 times 3 are 24, and *seven* (written in the rem.) make 31 (a number whose unit figure is *equal* to the right hand figure of the partial dividend); 8 times 2 are 16, and 3 (tens) from 31, are 19, and *one* (written in the rem.) makes 20. Final remainder, 17.

Find the number of Bushels in a car load of Corn weighing 20580 lbs.

We say 56 in 205, 3 times; 3 times 6 are 18, and *seven* (written in the rem.) make 25; 3 times 5 and 2 (tens) are 17, and *three* (written in the rem.) make 20.	56) 20580 (367½ bu. Ans. 378 420 28 *Rem.* $\frac{28}{56} = \frac{1}{2}$.

To the whole remainder 37, we bring down the 8, making 378. 56 in 378, 6 times; 6 times 6 are 36, and *two*, make 38; 6 times 5 and 3 (tens) are 33, and *four*, make 37. To the remainder 42, we annex the 0; 56 in 420, 7 times; 7 times 6 are 42, and *eight*, make 50; 7 times 5 and 5 (tens) are 40, and *two*, make 42. Final remainder, 28—which equals ½ bu.

1728 cubic inches make a cubic foot: how many cu. ft. in 233280 cu. in.?

1728 in 2332, 1 time, and 604 over; in 6048, 3 times, and 864 over; in 8640, 5 times—no rem.	1728) 233280 (135 cu. ft. Ans. 6048 8640

When the divisor is a mixed number, *write the fraction decimally, and annex ciphers to the dividend, till it has as many decimal places as the divisor; then proceed as in whole numbers.*

31½ gallons make a barrel: how many bbls. in 2394 gals.?	31.5) 2394.0 (76 bbls. Ans. 1890

If there is a remainder after the figures of the dividend are exhausted, *write a decimal point in the quotient, annex ciphers to the remainders, and carry the division on to two or more decimal places.*

24¾ cu. ft. of masonry make a perch: how many perches in 12583 cu. ft.?

```
24.75) 12583.00 (508.40 + p. Ans.
          208 00
           10 000
             1000
```

365¼ days make one year: how many years in 5918 days?

```
365.25) 5918.00 (16.2 + yrs. Ans.
         2265 50
           74 000
              950
```

When there are ciphers on the right of the divisor, see first example on page 33.

2000 lbs. make a ton: how many tons in a car load of coal weighing 21500 lbs.

```
2|000) 21|500
       ------
       10.75 tons. Ans.
```

160 square rods make an acre: how many acres in a field 83 rods long, and 72 rods wide?

```
         8 3
         7 2*
        -----
16|0) 597|6 (37.35  A. Ans.
      117
        56
         80
```

Having to some extent illustrated the principles involved in abbreviated multiplication and division, we will now present a series of special rules, or methods, based on these principles, for calculating the value of all kinds of Grain, Stock, Hay, Coal, Lumber, Merchandise, and particularly for computing Interest, and other problems in Percentage; also methods for ascertaining the capacity of Granaries, Corn Cribs, Cisterns; for finding the contents of Lumber, Land, etc.—all of which are specially adapted to the use of Farmers and Business men.

A comparison of these methods with those in general use, will readily convince any one of their simplicity, brevity, and practical superiority—results being usually obtained with about one-third the figures and mental labor required by the ordinary methods; and, besides, the tedious and much dreaded operations in fractions are easily and successfully surmounted.

*See "Simultaneous Multiplication," page 75.

GRAIN, HAY, COAL, ETC.

A simple, short, and practical method for finding the accurate value of articles sold by the *bushel* or *ton*, without involving fractions, even if the given terms are mixed numbers.

RULE.—*Write the number of lbs. to the bushel or ton, for the first term, the price for the second, and the weight for the third. Write common fractions decimally.*

Divide the second term by the first, and set the quotient, in REVERSED *order, under the third term.*

Multiply (by short method) *the third term by this quotient, point off two places from the right of the product, and the result will be the answer in dollars and cents.*

NOTE.—After the terms are stated, compare the second term with the first: if it is equal to, or greater than the first term, the first quotient figure must fall under the units' place of the third term; if smaller, or, in other words, if the price is *less* than a *cent per lb.*, the first quotient figure must fall under the *tens'* place of the third term.

Grain.

EXAMPLES.—Find the value of a load of Wheat weighing 2967 lbs., at $1.29\frac{3}{4}$ per bu.

Lbs. to bu.	Price per bu.	Weight.	
6|0	1.29-75	|2 9 6|7	
		5|2 6 1|2*	
		5 9.3 4	val. at 2 dol. per cwt.
		2.9 7	" " 1 dime " "
		1.7 8	" " 6 cts. " "
		6	" " 2 mills " "
		1	" " ½ " " "
		Ans. $6 4.1 6	

Having compared the 1st and 2d terms, and ascertained that the price is over a cent per lb., we cut off the 0 in the 1st term and say, 6 in 12, 2 times, which we set under the units (7) of the third term; then proceed: 6 in 9, 1 time, 3 over; in 37, 6 times, 1 over; in 15, 2 times, 3 over (assume a 0 joined to the 3); in 30, 5 times, which brings the quotient (multiplier) *one* place to the *left* of the 3d term (multiplicand). See note 1, page 5.

We now multiply (by short method) the 3d term by this

*The quotient, 52612, is the value per lb. or 100 lbs., in *reversed* order, at the rate of $1.29¾ per bu. or 60 lbs. The actual price per cwt. is $2.1625, or $2.16¼.

quotient, point off two figures from the right of the product, and the result is the answer, correct within a fractional part of a cent.

Find the cost of 2435 lbs. of Corn (in the ear), at 28 cts. per bu. of 70 lbs.

Here we say 70 in 28, no time, and write a 0 in the quotient, then cut off the 0 in the divisor and proceed; 7 in 28, 4 times, which terminates the division. We then multiply and point off as before.

70	.28	2 4 3 5
		4 0
		Ans. $9.7 4

Find the value of a car load of Corn (shelled), at $45\frac{3}{8}$ cts. per bu. Weight 21655 lbs.

Here we say 56 in 45, no time, write a 0 in the quotient, then proceed: 56 in 453, 8 times; 8 times 6 are 48, and *five* (written in the rem.) make 53; 8 times 5 and 5 (tens) are 45 (no rem.). We now bring down the 7 and say, 56 in 57, 1 time and 1 over; bring down the 5; 56 in 15, no time, write a 0 in the quotient, annex a 0 to the 15; 56 in 150, 2 times; 2 times 6 are 12, and *eight* (written in the rem.) make 20; 2 times 5 and 2 (tens) are 12, and *three* (written in the rem.) make 15. To the rem., 38, annex a 0; 56 in 380, 6 times, which is the last figure required in the quotient.

56	.45-375	2 1 6 5 5
	57	6 2 0 1 8 0
	150	1 7 3 2 4
	380	2 1 7
	Rem. 44	5
		Ans. $1 7 5.4 6

The equivalent decimal of $\frac{3}{8}$ is .375. See and study the table on page 28.

What will 8960 lbs. of Barley cost, at $98\frac{1}{2}$ cts. per bu.?

Say 48 in 98, 2 times, 2 over; in 25, no time; in 250, 5 times, 10 over; in 100, 2 times, 4 over; in 40, no time.

48	.98-5	8 9 6 0
	2 50	0 2 5 0 2
	100	1 7 9 2 0
	40	4 4 8
		1 8
		Ans. $1 8 3.8 6

How much will 3125 lbs. of Oats cost, at 34 cts. per bu. of 32 lbs.?

32	.34	3 1 2 5
	200	5 2 6 0 1
	80	3 1 2 5
	160	1 8 7
		8
		Ans. $3 3.2 0

Hay, Coal, Etc,

Find the value of a load of Hay weighing 1765 lbs., at $13.75 per ton.

Here we say 2000 in 1375, no time, write a 0 in the quotient, then cut off the ciphers in the divisor and proceed: 2 in 13, 6 times, 1 over; in 17, 8 times, 1 over; in 15, 7 times, 1 over; in 10, 5 times.

```
2|000      13.75        1765
                       57860
                       -----
                        1059
                         141
                          13
                 Ans. $12.13
```

Find the cost of 3270 lbs. of Coal, at $4.60 per ton.

2000 in 460, no time; 2 in 4, 2 times; 2 in 6, 3 times.

```
2|000       4.60        3270
                         320
                        ----
                         654
                          98
                  Ans. $7.52
```

Find the value of a lot of Straw weighing 9350 lbs., at $1.75 per ton.

```
2|000       1.75        9350
                       57800
                       -----
                         748
                          70
                  Ans. $8.18
```

NOTE.—When the price is *less* than $2 per ton, or a mill per lb., the first quotient figure must fall under the *hundredths'* place of the multiplicand.

STOCK, LUMBER, MERCHANDISE, ETC.

An easy, rapid, and simple method for finding the value of articles sold by the unit, hundred, or thousand, when one or both terms contain common or decimal fractions.

RULE.—*Write one of the terms, in* REVERSED *order, for the multiplier, so that the units of the one, and the hundredths or cents' order* (at the price per unit) *of the other, will fall in the same column.*

Multiply (by short method) *and point off two figures—the result will be dollars and cents, or units and hundredths.*

NOTES.—1. Although immaterial, it is usually most convenient to use the cost or price for the multiplier.

2. When the multiplier extends to the right of the multiplicand, assume or annex ciphers to the latter.

Stock.

EXAMPLES.—Find the value of 1385 lbs. of Pork, at $4.67½ per cwt.

$4 per 100 is 4 cts. per unit; hence the 4 must fall under the units (5) in the multiplicand, and the other figures, in *reversed* order, to the left of it.

```
      1 3 8 5
      5 7 6-4
      5 5.4 0  value at 4 dol. per cwt.
        8.3 1    "     " 6 dimes "    "
         .9 7    "     " 7 cts.  "    "
            7    "     " 5 mills "    "
Ans. $6 4.7 5
```

Find the value of a lot of fat Hogs weighing 8569 lbs., at $6.43¾ per cwt.

```
        8 5 6 9
      5 7 3 4-6
      5 1 4 1 4
        3 4 2 8
          2 5 7
            6 4
Ans. $5 5 1.6 3
```

What is the value of a Cow weighing 972 lbs., at 5¾ cts. per lb.?

```
          9 7 2
          5 7-5
        4 8 6 0
          6 8 0
            4 9
Ans. $5 5.8 9
```

Lumber.

Find the cost of a car load of Lumber, 4565 ft., at $21.62½ per thousand.

$20 per 1000 is $2 per 100, or 2 cts. per *unit*; hence the 2 must fall under the units.

```
        4 5 6 5
      5 2 6 1-2
        9 1 3 0
          4 5 7
          2 7 4
            1 1
Ans. $9 8.7 2
```

Find the cost of 683 ft. of common Lumber, at $16.50 per thousand.

```
          6 8 3
          5 6-1
          6 8 3
          4 1 0
            3 4
Ans. $1 1.2 7
```

Find the cost of 7750 Shingles, at \$4.87½ per 1000.

\$4 per 1000 is 40 cts. per 100, or 4 mills per *unit*. The order of *mills* must fall under the *tens*.

```
     7750
    5784
    ----
    3100
     620
      58
Ans. $37.78
```

Freight.

Find the Freight on a car load of Grain weighing 19875 lbs., at 34¼ cts. per cwt.

30 cts. per 100 is 3 mills per unit.

```
    19875
      543
     5963
      795
       99
Ans. $68.57
```

Find the Freight on a car load of Lumber, 5680 ft., at \$4.75 per 1000.

\$4 per 1000 is 4 mills per unit.

```
     5680
     574
    ----
    2272
     398
      28
Ans. $26.98
```

Merchandise.

Find the cost of 15 lbs. and 5 oz. of Butter, at 18¾ cts. per lb.

5 oz. or $\frac{5}{16}$ of a lb. is .31+, decimally. (See table, page 28.)
18 cts. is 1 dime and 8 cts.; hence the 8 must fall under the units. (5.)
Be careful to get the *cents'* order or *hundredths'* place of one term, and the *units* of the other, under each other.

```
     15.31
    57-81
    -----
      153
      122
       12
Ans. $2.87
```

Find the cost of 17½ doz. of Eggs, at 12½ cts. per doz.

The 7 occupies the units' place in the one term; 2, the cents' place in the other.

```
     17.5
     5-21
      175
       35
        9
Ans. $2.19
```

Find the cost of 37¾ yds. of Muslin, at 9¾ cts. per yard.

```
     37.75
    57-9
    -----
     340*
      28
Ans. $3.68
```

* See notes, page 37.

Find the value of a sack of Coffee weighing $216\frac{1}{2}$ lbs., at $23\frac{5}{8}$ cts. per lb.

$\frac{5}{8}$ is .625 decimally. (See table, p. 28.)

216.5
526-32
4330
650
130
5
Ans. $51.15

Find the cost of 48 lbs. of Sugar, at $13\frac{3}{4}$ cts. per lb.

In this and the next example, *reverse*, and write the *quantity* for the multiplier, setting the units under the cents' order.

.13-75
84
550
110
Ans. $6.60

Find the cost of $3\frac{1}{2}$ yds. of Cloth, at $\$1.16\frac{2}{3}$ per yd.

$\frac{2}{3}$ is .66 +, decimally. (See table, p. 28.)

$1.16-66
5.3
350
58
Ans. $4.08

Find the value of $26\frac{1}{2}$ bu. of Potatoes, at $1.05 per bu.

When the multiplier extends to the right, annex a 0 to the multiplicand.

26.50
50.1
2650
133
Ans. $27.83

NOTE.—When the answer is required correct to lower denominations, *write the multiplier further to the right, and point off from the product as many more places*, as will be illustrated in the following examples.

Find the cost of $8\frac{1}{2}$ lbs. of Nails, at $6\frac{1}{2}$ cts. per lb.

8.5
5-6
Ans. 55 cts.

8.5
5-6
510
43
Ans. 553 mills.

Find the cost of 34 lbs. and 13 oz. of Feathers, at $41\frac{2}{3}$ cts. per lb.

$\frac{13}{16}$ is .81 +, decimally. (See table, p. 28.)

34.81
66-14
1392
35
23*
Ans. $14.50

34.81
666-14
13924
348
209
23*
Ans. $14.504

*See notes, page 37.

PERCENTAGE.

Per cent. means on or by the hundred. Thus 1 per cent., denotes 1 out of a hundred, or 1 *hundredth;* 5 per cent. of a number means 5 hundredths of it.

The character %, is usually written instead of the word per cent.

How much is 6 % of $100?

	$1 0 0	= the	Base.
	6	" "	Rate %.
Ans.	$6.0 0	" "	Percentage.

The student should be careful to discriminate between percentage and product. The percentage is always the *hundredth* part of the product. Thus, 6 *times* $100 are $600, while 6 *per cent.* of $100 is $6.

Hence to find the percentage on any number;

Multiply the given number or base, by the rate per cent., and from the product point off two more decimal places than there are decimal places in the multiplicand; that is, divide the product by 100.

EXAMPLES.—Bought a lot of Hogs for $285, and sold them at 10 % profit: how much did I gain by the transaction?

	$2 8 5
	1 0
Ans.	$2 8.5 0

A merchant who failed in business, was able to pay 37 cts. on the dollar, or 37 %: what did A receive, who was a creditor to the amount of $2345?

	$2 3 4 5
	3 7
	1 6 4 1 5
	7 0 3 5
Ans.	$8 6 7.6 5

What is the commission, for selling $542 worth of property at 1½ %?

	$5 4 2
	1 ½
	5 4 2
	2 7 1
Ans.	$8.1 3

The following is a short method for finding the percentage when the given rate is a mixed number.

RULE.—*Write the fraction decimally; reverse and write the units of the rate % under the units, or dollars' place of the base, or vice versa.*

Multiply (by short method) *and point off two decimal places.*

I insured my house for \$965, at $2\frac{3}{4}$ %: what was the premium?

\$965
57·2
1930
676
48
Ans. \$26.54

What will be the commission for selling goods to the amount of \$6439.75, at $3\frac{3}{4}$ %?

\$6439.75
57·3
19319
4508*
322
Ans. \$241.49

A Railroad Company declares a dividend of $13\frac{5}{8}$ %: what will A receive, who owns \$3500 worth of stock?

13.625
0053
40875
6813
Ans. \$476.88

The principal application of Percentage is computing Interest, in which the element of Time is involved.

COMPUTING TIME.

To find the time between two dates, in years, months, and days: *Set the earlier date under the later, and subtract. Write the numbers of the months instead of their names.*

EXAMPLES.—Find the time from Jan. 6 to Sept. 18.

	Mo.		Da.	
	9		18	
	1		6	
Ans.	8	*mo.*	12	*da.*

We set down 9 and 18 for Sept. 18th, it being the 9th month; under this we write the earlier date, 1 and 6 for Jan. 6th, and then subtract.

Find the time from Oct. 20, 1871, to April 15, 1873.

*See notes, page 37.

April is the 4th month, Oct. is the 10th. We can not take 20 da. from 15 da.; we therefore conceive 30 da. (1 mo.) added to the 15 da., making 45; then say 20 from 45 leaves 25.

	Years.	Months.	Days.
	1873	4	15
	1871	10	20
Ans.	1 *yr.*	5 *mo.*	25 *da.*

Again, we can not subtract 10 mo. from 3 mo. (1 having been reduced to da.); hence we imagine 12 mo. (1 yr.) added to the 3 mo., and say 10 from 15 leaves 5. Finally we say, 1 year from 2 years (1 having been reduced to mo.) leaves 1 year.

INTEREST.

Interest is a percentage paid for the use of money.

Principal is the sum for the use of which int. is paid.

Rate per cent. is the sum paid on the hundred.

Per annum means by the year.

Amount is the interest and principal added together.

An easy, short, and simple method for finding the interest on any sum, for any time, at any rate per cent.

RULE.—*Write the whole number of months, with the order of its figures* REVERSED, *so that its units will fall under the units, or dollars' place of the principal.*

Divide the number of days by 3, *and write the quotient, in* REVERSED *order, to the left of the months.*

Multiply (by short method) *and point off two places from the product—the result will be the interest at* 12 *per cent.*

To obtain the interest at other rates by this method, *first find it at* 12 *per cent; then,*

For 10 %, *divide* IT *by* 6 *and subtract the quotient from the dividend.*
" 9 " " " " 4 " " " " " " "
" 8 " " " " 3 " " " " " " "
" 6 " " " " 2
" 4 " " " " 3

To find it at any other rate, *divide by* 12, *which gives the interest at* 1 %, *then multiply this quotient by the given rate.*

EXAMPLES.—Find the interest of $247.83 for 2 years 7 months and 13 days, at 12 %.

2 years and 7 mo. are 31 mo. We *reverse* this number, and write it so that its unit figure (1) will fall under the *units* (7), and the 3 under the *tenths* (8) of the principal. We then say, 3 in 13 (the number of days) 4 times, 1 over; we write the 4 under the principal, to the left of the months, and conceiving a 0 added to the 1 (rem.) we proceed 3 in 10, 3 times, which we set to the left of the 4; and thus we continue the division till the multiplier extends *one* place to the left of the multiplicand.

```
      $247.83
      334·13
      ------
       7435
        248
         99
          8
      ------
Ans.  $77.90
```

We then multiply (by short method) and point off two figures from the product, and the result is the int. at 12 %.

Find the interest of $86.50 for 5 mo. and 23 da., at 6 %.

We set the 5 mo. under the units, or dollars' place (6), then say 3 in 23, 7 times, 2 over; in 20, 6 times.

To obtain the int. at 6 %, *divide the int., at 12 % by 2.*

```
      $86.50
       67·5
      ------
       433
        65
      ------
   2) 4.98     int. at 12 %.
 Ans. $2.49     "   "   6 "
```

Find the int. of $165 for 1 yr. 4 mo. 12 da., at 6 %.

1 yr. and 4 mo. are 16 mo. Reverse, and write the 6 under the units (5), then say 3 in 12, 4 times.

```
       $165.00
          4·61
       -------
         1650
          990
           66
       -------
    2) 27.06     int. at 12 %.
Ans.   $13.53     "   "   6 "
```

Find the int. of $357 for 3 yrs. 7 mo. and 21 da., at 10 %.

3 yrs. and 7 mo. are 43 mo. *Reverse,* and write so that the 3 (units) will fall under the 7 (units); then say 3 in 21, 7 times.

```
       $357.00
          7·34
      --------
        14280
         1071
          250
      --------
    6) 156.01     int. at 12 %.
        26.00      "   "   2 "
      --------
Ans. $130.01      "   "  10 "
```

To obtain the interest at 10 %, *divide the interest at 12 % by 6—the quotient will be the interest at 2 %, which deducted from 12 % leaves 10 %.*

Find the interest of $617.50 for 26 days, at 10 %.

There being no months here, we write a 0 under the units, then say 3 in 26, 8 times, 2 over; in 20, 6 times, etc.

$617.50	
668-0	
494	
41	
6) 5.35	int. at 12 %.
.89	" " 2 "
Ans. $4.46	" " 10 "

Find the interest of $200 for 4 months 2 days, at 8 %.

Set the 4 months under the units, then say 3 in 2, no time; write a 0 next to the 4, and proceed: 3 in 20, 6 times, 2 over, etc.

$200.00	
660-4	
8.00	int. for 4 mo.
.13	" " 2 da.
3) 8.13	" at 12 %.
2.71	" " 4 "
Ans. $5.42	" " 8 "

To obtain the interest at 8 %, *divide the interest at 12 % by 3—the quotient will be the interest at 4 %, which deducted from 12 %, leaves 8 %.*

Find the interest of $90 for 10 months 12 days, at 7 %.

Set the 0 of the 10 (mo.) under the units, then say 3 in 12, 4 times.

$90.00	
4-01	
9.00	int. for 10 mo.
.36	" " 12 da.
12) 9.36	" at 12 %.
.78	" " 1 "
7	
Ans. $5.46	" " 7 "

To obtain the interest at 7 %, *divide the interest at 12 % by 12—the quotient will be the interest at 1 %, which multiplied by 7, gives it at 7 %.*

NOTE.—The lowest figure will not always be correct; to obviate this, first multiply the interest at 12 per cent. by 7, then divide by 12.

Find the amount of $58.75 for 1 year 8 months 19 days, at 10 %.

1 year and 8 months are 20 months. Set the 0 under the units, then say 3 in 19, 6 times, 1 over; in 10, 3 times.

$58.75	
360-2	
1175	
37	
6) 12.12	int. at 12 %.
2.02	" " 2 "
10.10	" " 10 "
58.75	*Principal.*
Ans. $68.85	*Amount.*

To obtain the amount, *add the principal to the interest.*

To find the interest of $10, $100, $1000, etc., for any time at 12 %, *set down the whole number of months and annex ⅓ of the days; then place the decimal point correctly.* Thus,

The int. of 1 dol. for 1 yr. 4mo. 25 da., at 12 %, is		.16|833
" " " 10 " " " " " " " "		$1.68|33
" " " 100 " " " " " " " "		$16.83|3
" " " 1000 " " " " " " " "		$168.33|

It is evident that the interest of $30 is 3 times that of $10; of $400, 4 times that of $100, etc., for the same time and rate. Thus, in the above illustration, the interest of $100 is $16.83⅓; hence, for $300, it would evidently be 3 times $16.83⅓ = $50.50.

Interest—Accurate Method.

The preceding method, and nearly all others in general use, do not give the interest strictly correct for the months and days; 30 days being considered a month, and, consequently, 360 days a year, and interest reckoned accordingly for a fractional part of a year, is usually found too large.

The *true* interest of $1200, at 10 %, for 31 days, is $10.19; for 30 days, $9.86; and for 28 days, $9.20; while by the ordinary methods it would be just $10 a month, whether it consisted of 28, 29, 30, or 31 days.

This false principle of computing interest (for months and days) on the basis of 360 days to the year, instead of 365, has become customary in the United States—the State of New York excepted—*for convenience' sake only;* it being considerably easier to calculate by this than by the true basis.

We will now present a method for finding the *accurate* interest on any sum, for any time, at any rate per cent., which we claim to be entirely original, and unequaled for simplicity and brevity.

RULE.—*Find the time in years and days, multiply the number of days by* 472 (by short method), *and to the product prefix the years, if any.*

Reverse, and write this number with its cents' order under the units, or dollars' place of the principal.

Multiply again (by short method), *and point off two decimal places from the product—the result will be the accurate interest at* 10 *per cent.*

To find the interest at any other rate % by this method, *multiply the interest at* 10 % *by the given rate, and point off* THREE *places from the last product.*

NOTES.—1. Memorize the number 472; this is the interest of $1 for 1 day at 10 per cent., with the order of its figures reversed and the ciphers omitted; the real number being $.00-0274, or more accurately, $.00-02739726 + .

2. When the first product (the number of days multiplied by 472) contains less than *three* places, *prefix* ciphers to supply the deficiency. This, however, happens only when the number of days is *less* than 37.

3. When the principal is large, and great accuracy required, *annex* a 0 to the number of days, and proceed as usual—the product must then contain *four* places.

The exact number of days from one date to another is readily found by the Time Table on page 24; or by the following method.

Find the true number of days from January 10 to May 15.

After the 10th there are 21 days in Jan., which, with the 15 in May, added to the days in the intervening months, gives the exact number.

In Jan.	21	da.
" Feb.	28	"
" Mar.	31	"
" Apr.	30	"
" May	15	"
Ans.	125	"

EXAMPLES.—Find the accurate interest of $354.56, at 10 %, from March 11 to December 24 = 288 days.

We first multiply the number of days (288) by 472; the product (789) is the interest of $1 for 288 days, at 10 %, namely, 7 cts., 8.9 mills. Now it is evident that of $354, the interest is 354 times that of $1. We therefore write the interest of $1 in *reversed* order under the principal, so that its *left hand* figure or *cents'* order (7) will fall under the *units*, or *dollars'* place (4).

```
       $354.56          288 da.
         98-7           472
       ------           ---
        2482            576
         283            202
          32             11
  Ans. $27.97           ---
                        7-89
```

We then multiply again (by short method), point off two decimal places, and the result is the accurate interest at 10 %.

Find the true interest of $75 at 10 %, from June 1, 1872, to July 5, 1873 = 1 year and 34 days.

Here the first product contains only two places—9.3 mills; hence we must write a 0 in the cents' place.

```
       $75.00       1 yr.    34 da.
       39-01                 472
       -----                 ---
        750                   68
         70                   25
  Ans. $8.20                 ----
                             10-93
```

Now the interest of $1 for 1 year at 10 % is just 1 dime; we therefore prefix 1 (or whatever the number of years may be) to the interest for the days, and the result is the interest of $1 for the entire time. We now *reverse* this number and set it under the principal, being careful to get the *cents'* order, under the *units*, or *dollars'* place, then multiply and point off as before.

Find the true int. of $167.85 for 2 yrs. and 3 da., at 6 %.

Here we prefix two ciphers to the first product 8 (.8 of a mill) before prefixing the 2 (yrs.), in order to bring the significant figures into their respective places.

$1 6 7.8 5		2 yrs. 3 da.
8 0-0 2		4 7 2
3 3.5 7	int. for 2 yrs.	2 0-0 8
.1 3	" " 3 da.	
3 3.7 0	" at 10 %.	
6		
Ans. $2 0.2 2 0	" " 6 "	

To obtain the interest at 6 %, *multiply the interest at* 10 % *by* 6, *and point off* THREE *places from the last product.*

Find the accurate interest of $936.75 for 293 days, at 5 %.

When the principal is large and great accuracy required, we *annex* a 0 to the number of days, then proceed as usual.

$9 3 6.7 5		2 9 3.0 da.
8 2 0-8		4 7 2
7 4 9 4		5 8 6 0
2 6		2 0 5 1
2) 7 5.2 0	int. at 10 %.	1 1 7
Ans. $3 7.6 0	" " 5 "	8-0 2 8

To obtain the interest at 5 %, *divide the interest at* 10 % *by* 2.

Find the true interest of $2683.43 for 1 year and 33 days, at 7 %.

$2 6 8 3.4 3		1 yr. 3 3.0 da.
4 0 9-0 1		4 7 2
2 6 8 3 4		6 6 0
2 4 1 5		2 3 1
1 1*		1 3
2 9 2.6 0	int. at 10 %.	1 0-9 0 4
7		
Ans. $2 0 4.8 2 0	" " 7 "	

When a 0 is annexed to the number of days, the product must contain *four* places. Hence a cipher must be prefixed here before we prefix the 1 year.

* See notes, page 37.

Find the amount of $67.92 for 343 days, at 10 %.

To obtain the amount, *add the principal to the interest.*

```
      $67.92              343 da.
       04-9               472
       ----               ---
       611                686
        27                240
       ----                14
       6.38 Interest.     ---
      67.92 Principal.    9-40
      ------
Ans. $74.30 Amount.
```

To find the true interest of $10, $100, $1000, etc., for any time, at 10 %, *annex a 0 to the number of days, multiply by* 472 (short method), *and to the product prefix the years, if any; then place the decimal point correctly.* Thus, the true interest

```
1 yr.                                        93.0 da.
                                             472
                                             ----
                                             1860
                                              651
                                               37
                                             -----
Of    1 dol. for 1 yr. 93 da., at 10 %, is  .12548
"    10  "    "   "    "       "    "   "  $1.2548
"   100  "    "   "    "       "    "   "  $12.548
"  1000  "    "   "    "       "    "   "  $125.48
```

It is obvious that multiplying the int. of $100 by 4, gives it for $400; multiplying the int. of $1000 by 3, gives it for $3000, etc. Thus, in the above illustration, the int. of $1000 is $125.48: evidently, for $2000 it would be *twice*—for $3000, *three* times—$125.48., and so on.

PARTIAL PAYMENTS.

A ***Partial Payment*** is the payment of a part of the amount due on a note or bond.

The following (called the Common, Vermont, or Merchants') Rule for computing interest on notes where partial payments have been made, is simple, easily comprehended, and extensively used by Merchants and Farmers.

It is based on the principle, that as the creditor receives interest on money loaned, so he should pay interest on money received before it becomes due. It is the only rule that does justice to the debtor when payments have been made at short intervals. When, however, the time from date to settlement extends into years, it favors the debtor, as it generally should if the rule is to favor any one—no interest being paid till the time of settlement.

RULE.—*Find the amount of the principal from the time it began to draw interest to the day of settlement.*

Find the interest on each payment from the time it was made to the day of settlement.

Subtract the sum of the payments and interest thereon from the amount of the principal—the remainder will be the sum due on settlement.

$100. BLOOMINGTON, ILL., Jan. 1, 1872.

One day after date, I promise to pay to Charles Jones, or order, One Hundred Dollars, for value received, with interest at ten per cent. per annum. JOHN SMITH.

Indorsements:

May 1, 1872, Received Sixty Dollars.
Sept. 1, 1872, Received Thirty Dollars.

How much was due at the time of settlement, Jan. 1, 1873?

Principal,		$100
Interest from Jan. 1, '72, to Jan. 1, '73 (1 yr.),		10
Amount of note Jan. 1, '73,		$110
First payment, made May 1, '72,	$60	
Int. on same to Jan. 1, 73 (8 mo.),	4	
Second payment, made Sept. 1, '72,	30	
Int. on same to Jan. 1, '73 (4 mo.),	1	
Amount of payments and interest thereon,	$95	95
Balance due Jan. 1, 1873,		$15

$834.75. CHICAGO, ILL., May 14, 1870.

On or before the first of January, 1873, we, or either of us, promise to pay to Robert Brown, or bearer, Eight Hundred, Thirty-Four and $\frac{75}{100}$ Dollars, for value received, with six per cent. interest from date.

WILLIAM WHITE.
GEORGE GREEN.

Indorsements:

Oct. 20, 1871, Received $217.45.
Feb. 6, 1871, Received $475.00.
July 17, 1872, Received $124.30.

How much remained due Jan. 1, 1873?

Principal,..		$834.75
Interest from May 14, '70, to Jan. 1, '73, (2 yrs. 7 mo. 17 days),..............................		131.75
Amount of note, Jan. 1, '73,..........................		$966.50
First payment, made Feb. 6, '71,..........	$475.00	
Int. to Jan. 1, '73, (1 yr. 10 mo. 25 da.),	54.23	
Second payment, made Oct. 20, '71,......	217.45	
Int. to Jan. 1, '73 (1 yr. 2 mo. 11 da.),..	15.62	
Third payment, made July 17, '72	124.30	
Int. to Jan. 1, '73 (5 mo. 14 da.),.........	3.40	
Amount of payments and interest thereon..	$890.00	890.00
Balance due Jan. 1, 1873,..............................		$76.50

$495. NEW YORK, March 15, 1872.

Twelve months after date, we promise to pay to the order of David Pope & Son, Four Hundred and Ninety-Five Dollars, for value received, with ten per cent. interest. Payable at the Lafayette Bank.

KING, HALE & CO.

Indorsements:

Sept. 22, 1872, Received $375.00
April 11, 1873, Received $107.50.

What remained due, July 4, 1873?

The interest on this note is computed by the accurate method.

Principal,...		$495.00
Int. from Mar. 15, '72, to July 4, '73 (1 yr. 111 da.),		64.55
Amount of note, July 4, '73,................................		$559.55
First payment, made Sept. 22, '72,.......	$375.00	
Int. to July 4, '73 (285 da.),...............	29.29	
Second payment, made April 11, '73,....	107.50	
Int. to July 4, '73 (84 da.),.................	2.47	
Amount of payments and int. thereon,....	$514.26	514.26
Balance due July 4, 1873..................................		$45.29

DISCOUNT AND PRESENT WORTH.

Discount is an allowance made for the payment of a debt before it is due.

The ***Present Worth*** of a note, due at a future time

without interest, is such a sum which, being put at interest now, will amount to the given debt when it becomes due. Thus, $100 is the *present worth* of $110 due one year hence without interest, discounted at 10 %; for $100 at 10 % will amount to $110 in that time—$10 being the *discount.*

To find the *true* discount and present worth of a note or debt.

RULE.—*Divide the given debt by the amount of $1 for the given time and rate, the quotient will be the* PRESENT WORTH.

Subtract the present worth from the given debt—the remainder will be the true DISCOUNT.

EXAMPLES.—Find the present worth, and true discount of $156.75, due 1 yr. hence, at 10 %.

The amount of $1 for 1 yr. at 10 % is $1.10. We divide $156.75 by $1.10; the quotient is the present worth, which subtracted from the given sum, leaves the true discount.

```
1.10) 156.75 (142.5 Pres. Worth.
       467
        275
         55

    $156.75 Sum or Debt.
    $142.50 Present Worth.
     $14.25 True Discount.
```

PROOF.—The Int. of $142.50 for 1 yr. at 10 % is $14.25, which added to the principal, or present worth, gives the amount or debt $156.75.

Bought a Horse for $130 on 8 mo. credit. What would be the present worth of the debt, discounted at 6 %?

The amount of $1 for 8 mo. at 6 % is $1.04.

```
1.04) 130.00 (125 dol. Ans.
       260
        520
```

Find the present worth and discount of $413.65, payable in 96 days, discounted at 10 %.

The int. of $1, at 10 % is readily found by multiplying the given No. of days by 472 (short method). The amount is $1.0263.

```
  96      1.0|2|6|3) 413.65 (403.05 P. W.
 472                  313*
 192                    5
  71
 263

          $413.65 Debt.
          $403.05 Present Worth.
           $10.60 Discount.
```

* See "Contracted Division," page 78.

BANK DISCOUNT.

Bank Discount is the simple interest of a note or debt, deducted from it in advance, or before it becomes due. When money is obtained at a bank, the interest for the specified time, and *three days more*—called "days of grace," is deducted from the sum or face of the note in advance, the remainder being called AVAILS or PROCEEDS.

RULE.—*Compute the interest on the face of the note at the given rate % for* THREE *days more than the specified time, the result will be the discount.*

Subtract the discount from the sum or face of the note, and the remainder will be the proceeds.

EXAMPLES.—What is the bank discount, and what are the proceeds of a note for $100, on 30 days time, at 10 % ?

By multiplying the No. of days (33) by 472 (short method), we obtain the accurate int. of $1 at 10%, which is 9 mills. Now, for $100 it must evidently be 100 times 9 mills—that is 90 cts.; which, deducted from $100, leaves the proceeds.

33 da.	$100.00	*Sum.*
472	.90	*Discount*
66	$99.10	*Proceeds.*
24		
0.90		

Find the bank discount on a note for $1000, payable in 90 days, at 10 %.

The given sum being large, we annex a 0 to the No. of days (93).

The product 2548 is the accurate int. of either 1, 10, 100, or 1000 dollars—dependent on where we place the decimal point. Now, it is readily perceived that the int. of $1000 for 93 days at 10 per %, must be more than $2.548, and that it can not be $254.8; consequently, it must be $25.48.

93.0	$1000.00	*Face of Note.*
472	25.48	*Discount.*
1860	$974.52	*Proceeds.*
651		
37		
Ans. 2548		

What is the bank discount on $546.87 for 70 days, at 10 % ?

The int. of $1 for 73 days at 10 %, is just 2 cts; it is then easily found for $546.87.

$546.87	73 da.
00.2	472
Ans. $10.94	146
	54
	2.00

Find the bank discount and proceeds of $97.68, due in 1 yr. 2 mo. 10 days, discounted at 6 %. Computed by the first rule for casting Interest.

```
    $97.68              14 mo. 13 days.
    3441
     977
     391*               $97.68 Sum.
      42                  7.05 Discount
 2) 14.10               $90.63 Proceeds.
    $7.05 int., at 6 %.
```

The difference between *true* and *bank* discount is insignificant for short periods of time, but increases in a fearful ratio as the time extends into years, as will be seen in the following illustration.

How much would I receive for a note of $1000, due in 10 years, without interest, if discounted at 10 % true discount, and how much if discounted at the same rate % by bank discount, not reckoning days of grace?

Ans. $500 by true discount.
Nothing by bank discount.

True Method.		Banker's Method.	
$2.00	Amount of $1 for 10 yrs.	$1000.	Face of Note.
2.00) 1000.00	Face of Note.	$1000.	Int. for 10 yrs.
Ans. $500	Present Worth.	0000.	Proceeds.

True discount is the interest on the *present worth* of a note, which is always *less* than its face.

Bank discount is the interest on the *face* of a note, and the interest deducted from it leaves the *proceeds*. Hence, whenever the interest equals the face of the note or debt, there will be no proceeds left; that is, any note without interest becomes worthless, when discounted by bankers' method, in the same time that it would double itself at the given rate %.

PROFIT AND LOSS.

Profit or Loss is the difference between the cost of an article and the amount received for it. The *Gain* or *Loss* is always estimated on the cost price.

* See notes on page 37.

To find the gain or loss, when the cost price and gain or loss per cent. are given.

RULE.—*Multiply the cost price by the gain or loss per cent., and from the product point off two more decimal places than there are decimals in the multiplicand—the result will be the gain or loss.*

To find the selling price, *the gain or loss is added to, or subtracted from the cost price.*

EXAMPLES.—Flour that cost $7.50 per bbl., was sold at 12 % profit: what was the gain, and what was the selling price per bbl?	$7.50 12 Ans. .90-00 Gain per bbl. 7.50 Cost price per bbl. Ans. $8.40 Selling price per bbl.
A wagon that cost $115 was sold at a discount of 20 %: what was the loss, and what was the selling price?	$115 20 Ans. $23.00 Loss. $115 Cost price. 23 Loss. Ans. $92 Selling price.
What must goods that cost 25 cts. per yd. be sold at so as to make 15 %.	.25 15 .0375 Gain per yd. .25 Cost price Ans. .28-75 = 28¾ cts. per yd.
I cleared 8¾ % on a lot of Hogs which cost me $625: what did I gain, and what did I get for the lot?	625 57-8 Gain % *reversed.* 5000 438 31 Ans. 54.69 Gain. 625.00 Cost price. Ans. $679.69 Selling price.

To find the gain or loss per cent. when the cost and selling price are given.

RULE.—*Find the difference between the cost and selling price, which will be the gain or loss.*

Annex two ciphers to the gain or loss, and divide it by the cost price—the result will be the gain or loss per cent.

Bought Wheat at $1.25 per bu. and sold it for $1.50: what per cent. did I make by the transaction?

$1.50	Selling price.
1.25	Cost price.
.25	Gain per bu.

1.25) 25.00 (20 %, Ans.

A merchant sold cloth at $3 per yd. that cost $3.60: what % did he lose?

$\frac{24}{36}$ reduced to its lowest terms, equals $\frac{2}{3}$.

$3.60	Cost price.
3.00	Selling price.
.60	Loss per yd.

3.6|0) 60.0|0 (16$\frac{2}{3}$ %, Ans.
240
24 *Rem.* $\frac{24}{36} = \frac{2}{3}$.

A grocer sells coffee at 25 cts. per lb. that cost him 22: what % does he make?

.25	Selling price.
.22	Cost price.

22) 3.00 (13$\frac{7}{11}$ %, Ans.
80
14 *Rem.* $\frac{14}{22} = \frac{7}{11}$.

A man paid $324 for a Team, and sold it again for $315.90: what % did he lose?

Instead of annexing two ciphers to the dividend (8.10), we omit those in the divisor (324.00).

$324.00
315.90

324) 8.10 (2$\frac{1}{2}$ %, Ans.
1.62 *Rem.*

$\frac{162}{324} = \frac{1}{2}$

A Farm that cost $4800, was sold for $5000: what % was gained by the transaction?

$5000
4800

48) 200 (4$\frac{1}{6}$ %, Ans.
8 *Rem.* $\frac{8}{48} = \frac{1}{6}$

GOLD AND CURRENCY.

Gold is usually represented as rising and falling, but being the standard of value, it does not vary. The variation is in the currency substituted for gold or specie; hence, when gold is said to be at a premium, the currency or circulating medium is made the standard, while it is in fact below par.

To change gold into currency. RULE.—*Multiply the given sum of gold by the price of gold.*

EXAMPLES.—How much currency can be obtained for $362.50 in gold, when gold is at 108 %, or $1.08?

We reverse and write the price with its cents' order under the units of the given sum, then multiply by short method and point off two figures.

```
       $3 6 2.5 0
           8 0.1
       ----------
         3 6 2 5 0
           2 9 0 0
       ----------
Ans.   $3 9 1.5 0
```

How much currency can be obtained for $85 in gold, it being at 112¾ %?

Here we reverse and take the sum or quantity for the multiplier, placing units under cents, or hundredths.

```
       $1.1 2-7 5
            5 8
       ---------
        9 0 2 0
          5 6 4
       ---------
Ans.   $9 5.8 4
```

To change Currency into Gold. RULE.—*Divide the amount in currency by the price of gold.*

How much gold can be obtained for $70.85 in greenbacks, gold being at 109 %, or $1.09?

```
1.0 9) 7 0.8 5 (6 5 dol. Ans.
         5 4 5
```

How much gold can be bought for $175 in currency, gold being at 113¾ %?

```
*1,.1|3|7|5) 1 7 5.0 0 (1 5 3.8 4 Ans.
               6 1 2 5
                 4 3 7
                   9 6
                     5
```

When gold is at a certain per cent. premium over currency, the discount on the currency is not the same as the premium on gold; thus, when gold is at 25 % premium, the corresponding discount on currency is but 20 %; and when gold is at 200 %, or 100 % premium, $1 currency is worth 50 cts. in gold; but when the discount on currency is 100 %, it is entirely worthless.

To find the corresponding value and discount on Currency when the premium or price of gold is given.

RULE.—*Annex two ciphers to* 100, *and divide it by the price of gold; the quotient will be the value* (in gold) *of* $1 *currency; and the difference between this sum and* 100, *will be the discount on currency.*

* See "Contracted Division," page 78.

When gold is 20 % premium, or at 120 %; what is the corresponding value and discount on currency?

$$\begin{array}{l} 12|0)\ 100.0|0\ (83\tfrac{1}{3}\text{ cts. Ans.} \\ \quad 40 \\ \quad\quad 4\ \textit{Rem.}\ \tfrac{4}{12} = \tfrac{1}{3}. \\ 100 \\ \underline{\ \ 83\tfrac{1}{3}} \\ \text{Ans. } 16\tfrac{2}{3}\ \%\ \text{Discount.} \end{array}$$

To find the corresponding price and premium on gold, when the value or discount on currency is known.

RULE.—*Annex two ciphers to* 100, *and divide it by the value* (in gold) *of* $1 *currency; the quotient will be the price of gold in currency, and the difference between this sum and* 100 *will be the premium.*

When the discount on currency is 25 %, or $1 currency is worth 75 cts. in gold; what is the corresponding price, and premium on gold?

$$\begin{array}{l} 75)\ 100.00\ (133\tfrac{1}{3}\text{ Price, Ans.} \\ \quad 250 \quad\quad 100 \\ \quad 250 \quad\quad \ \ 33\tfrac{1}{3}\text{ Premium, Ans.} \\ \quad\ \ 25\ \textit{Rem.} \\ \quad\ \ \tfrac{25}{75} = \tfrac{1}{3}. \end{array}$$

TABLE,

Showing the comparative value of Gold and Currency.

When the price of $1 Gold is (in Currency)	The Premium on Gold is	The Corresponding val. of $1 Currency is (in Gold)	The Discount on Currency is
101 cts.	1 %	$99\frac{1}{101}$ cts.	$\frac{100}{101}$ %
105 "	5 "	$95\frac{5}{21}$ "	$4\frac{16}{21}$ "
110 "	10 "	$90\frac{10}{11}$ "	$9\frac{1}{11}$ "
115 "	15 "	$86\frac{22}{23}$ "	$13\frac{1}{23}$ "
120 "	20 "	$83\frac{1}{3}$ "	$16\frac{2}{3}$ "
125 "	25 "	80 "	20 "
$133\frac{1}{3}$ "	$33\frac{1}{3}$ "	75 "	25 "
150 "	50 "	$66\frac{2}{3}$ "	$33\frac{1}{3}$ "
$166\frac{2}{3}$ "	$66\frac{2}{3}$ "	60 "	40 "
200 "	100 "	50 "	50 "
500 "	400 "	20 "	80 "
1000 "	900 "	10 "	90 "
10000 "	9900 "	1 "	99 "

PARTNERSHIP OR COMPANY BUSINESS.

A ***Partnership or Firm*** is an association of two or more persons, for the purpose of transacting business with an agreement to share the profits and losses proportionally.

Capital or Joint Stock is the amount of money or property used in the business.

Dividend is the amount of profit or loss apportioned to each partner.

To find each partner's share of the gain or loss.

RULE.—*Divide the whole gain or loss by the entire stock, the quotient will be the gain or loss per cent.*

Multiply each partner's stock by this per cent., the result will be each one's share of the gain or loss.

EXAMPLES.—Smith and Jones entered into partnership with a capital of $6000, of which Smith furnished $3500, and Jones $2500. They gain $600; what was each one's share of the gain?

6000) 600.00 (.10, or 10 cts. gain on the dollar.
$3500 × .10 = $350 Smith's share of the gain.
$2500 × .10 = $250 Jone's share of the gain.

A, B, and C, rented a farm for $960. They cleared above all expenses $456. What % did they gain on their money, and what was A's share who furnished $350?

9 6|0) 4 5 6.0|0 (.4 7 $\frac{1}{2}$ Ans.
7 2 0
4 8 *Rem.* $\frac{48}{96} = \frac{1}{2}$.
$350 × .47$\frac{1}{2}$ = $166.25 A.'s share.

Thompson, Clark & Co., have failed in business. Their liabilities or debts amount to $42650, and their assets or available property to $23884. How much can they pay on the dollar, and what dividend will Franklin Radford receive, whose claim is $750?

4 2 6 5|0) 2 3 8 8 4.0|0 (.5 6, or 56 cts. on the dollar
2 5 5 9 0
$7 5 0 × .5 6 = $4 2 0, Radford's share.

This is usually termed Bankruptcy, but is computed on the same principle as partnership.

Mike, Dick, and Patrick dug a ditch for $100. Mike worked 13, Dick 10, and Patrick 9 days. What wages did they make per day, and what was each one's share of the $100?

We divide the $100 by 32, the whole No. of days worked, the quotient will be the wages per day, which multiplied by the number of days that each one worked, will give each one's share.

```
32)100.00(3.125 wages per day.
     40
      80
      160

$3.125 × 13 = $40.62½ Mike's share.
$3.125 × 10 = $31.25  Dick's    "
$3.125 ×  9 = $28.12½ Patrick's "
```

LEVYING TAXES.

Taxes are assessments laid on property for the purpose of defraying public expenses.

To find the rate of taxation, the required tax and the value of the taxable property being known.

RULE.—*Annex ciphers to the number denoting the tax, and divide it by the number denoting the taxable property, the quotient will be rate of taxation.*

EXAMPLES.—In a certain school district, valued at $48350, it becomes necessary to levy a tax of $967 for school purposes. What will be the rate of taxation, and what will be Henry Harper's school tax, whose property is valued at $3765?

```
4835|0)967.0|0(.02, or 2 cts. on the dol.

         $3765 Harper's property.
           .02
Ans.    $75.30    "      School tax.
```

An iron bridge which cost $1353.75, was built by a township whose taxable property is valued at $386718. What will be the tax on the dollar, and what will be John Sherman's bridge tax whose property is valued at $7284?

```
386718)1353.7500(.0035+, or 3½ mills on the dol.
        1935960
           2370 Rem.

       $7284   Sherman's property.
        .003½
       21852
        3642
      $25.494 Sherman's bridge tax.
```

GROSS AND NET WEIGHT AND PRICE OF HOGS.

A short and simple method for finding the net weight, or price of Hogs, when the gross weight or price is given, and *vice versa*.

NOTE.—It is generally assumed that the gross weight of Hogs, *diminished* by 1-5 or 20 per cent. of itself gives the net weight, and the net weight *increased* by ¼ or 25 per cent. of itself, equals the gross weight.

To find the Net weight, or Gross price; *Multiply the given number by* .8 (tenths).

EXAMPLES.—A hog weighing 365 lbs. gross, will weigh 292 lbs. net; and Pork at $3.65 net, is equal to $2.92 gross.

365
.8
292.0

What will be the Net weight of a Hog that weighs 485 lbs. gross?

485
.8
Ans. 388.0 lbs.

To find the Gross weight or Net price; *Divide the given number by* .8 (tenths).

EXAMPLES.—A Hog weighing 348 lbs. net, weighed 435 lbs. gross; and Pork at $3.48 gross, is equal to $4.35 net.

.8) 348.0
435

$4.75 per cwt. for Hogs gross, is equal to what price net?

.8) 475.0
Ans. $5.93¾

MENSURATION.

Mensuration is the art of measuring surfaces, and determining the area and solid contents of geometrical figures or bodies.

We here present a series of short and simple methods for ascertaining the contents, or capacity of Granaries, Corncribs, Cisterns, Casks, etc.; also rules for measuring Lumber, Logs, Land, and numerous other things, all of which are of practical utility to Farmers, Merchants, and Mechanics.

Grain Measure.

To find the capacity of a Granary, Bin, or Wagon-bed.

RULE.—*Multiply* (by short method) *the number of cubic feet by 6308, and point off* ONE *decimal place—the result will be the correct answer in bushels and tenths of a bu.*

For only an approximate answer, *multiply the cu. ft. by 8, and point off one decimal place.*

EXAMPLES.—Find the capacity of a Granary 18 ft. long, 9 ft. wide, and 8 ft. high.

To obtain the number of cu. ft. we *multiply the length, width, and height together.*

```
18 × 9 × 8 = 1296 cu. ft.
             6308
            -----
            10368
               39
                7
            -----
     Ans. 1041.4 bu.
```

What is the capacity of a Bin 9 ft. long, 6 ft. wide, and $7\frac{1}{2}$ ft. deep?

```
9 × 6 × 7½ =  405 cu. ft.
             6308
             ----
             3240
               14
             ----
     Ans. 325.4 bu.
```

How much grain will a Wagon-bed hold that is 11 ft. long, 3 ft. wide, and 2 ft. deep?

```
11 × 3 × 2 =   66 cu. ft.
             *308
             ----
     Ans.  53.0 bu.
```

Find the contents a Wagon-bed 11 ft. 11 in. long, 3 ft. 1 in. wide, and 1 ft. 8 in. deep.

We write the inches decimally, thus, 11 in. or $\frac{11}{12}$ equals .91 + ft., $\frac{1}{12}$ = .08 + ft., 8 in. or $\frac{2}{3}$ = .66 + ft. See table, page 28.

```
  11.91 length.
   80.3 width reversed.
   ----
     37
   66.1 depth reversed.
   ----
     37
     24
   ----
     61 cu. ft
   *308
   ----
Ans. 49.0 bu.
```

To find the contents of a Corn-crib.

RULE.—*Multiply the number of cubic feet by 54, short method, or by $4\frac{1}{2}$ ordinary method, and point off* ONE *decimal place—the result will be the answer in bushels.*

* Here the 6 becomes superfluous, and hence is omitted.

EXAMPLES.—Find the contents of a Corn-crib 14 ft. long, 7 ft. wide, and 9 ft. high.

$$
\begin{array}{r}
14 \times 7 \times 9 = 882 \text{ cu. ft.} \\
54 \\
\hline
3528 \\
441 \\
\text{Ans. } 396.9 \text{ bu.}
\end{array}
$$

How many bu. will a crib hold that is 48 ft. long, $7\frac{1}{2}$ ft. wide, and $8\frac{1}{2}$ ft. high?

$$
\begin{array}{r}
48 \times 7\frac{1}{2} \times 8\frac{1}{2} = 3060 \text{ cu. ft.} \\
4\frac{1}{2} \\
\hline
12240 \\
1530 \\
\text{Ans. } 1377.0 \text{ bu.}
\end{array}
$$

Hay Measure.—About 500 cubic feet of well settled hay, or about 700 of new mown hay will make a ton.

NOTE.—The only *accurate* method to measure hay is to weigh it, since two quantities equal in bulk will never weigh alike. Any rule is simply an approximation.

Cistern, Tank, and Barrel Measure.

To find the contents of a Cistern or Tank.

RULE.—*Multiply the square of the mean diameter by the depth,* (all in feet) *and this product by* 5681 (short method), *and point off* ONE *decimal place—the result will be the contents in barrels of* $31\frac{1}{2}$ *gallons.*

EXAMPLES.—Find the contents of a Cistern whose mean diameter is 9 ft., and depth 10 ft.

$$
\begin{array}{r}
9 \times 9 \times 10 = 810 \\
5681 \\
810 \\
648 \\
53 \\
\text{Ans. } 151.1 \text{ bbl.}
\end{array}
$$

The square of a number is the product of that number multiplied by itself. Thus, the square of 9 (9 times 9) is 81; this multiplied by 10 makes 810, which we multiply by 5681, short method.

Find the contents of a Cistern 6 ft. in diameter, and $7\frac{1}{2}$ ft. deep.

$$
\begin{array}{r}
6 \times 6 \times 7\frac{1}{2} = 270 \\
5681 \\
\hline
270 \\
216 \\
17 \\
\text{Ans. } 50.3 \text{ bbl.}
\end{array}
$$

Find the capacity of a Tank 18 ft. in diameter, and 22 ft. deep.

```
18×18×22=7128
            5681
            ----
            7128
            5702
             428*
              36
            ----
Ans.  1329.4 bbl.
```

To find the contents of a Barrel or Cask.

RULE.—*Under the square of the mean diameter, write the length* (all in inches) *in* REVERSED *order so that its* UNITS *will fall under the* TENS; *multiply by short method, and this product again by* 430; *point off one decimal place, and the result will be the answer in wine gallons.*

EXAMPLES.—Find the contents of a Barrel, the mean diameter of which is 19 in., and the length 35 in. Also of a Cask whose mean diameter is 22½ in. and length 40½ in.

```
        Barrel.                          Cask.
19×19=361                    22½×22½=506
       53 length reversed.           5.04 length reversed.
     ----                            ----
     1083                            2024
      181                              25
     ----                            ----
     1264                            2049
      430                             430
     ----                            ----
      379                             615
       50                              82
     ----                            ----
Ans. 42.9 gal.                  Ans. 69.7 gal.
```

Lumber Measure.

To measure Boards. RULE.—*Multiply the length* (in feet) *by the width* (in inches) *and divide the product by* 12—*the result will be the contents in square feet.*

EXAMPLES.—How many sq. ft. in a board 18 ft. long, 10 in. wide?

```
18×10=180,   12)180
             Ans. 15 ft.
```

In a board 16 ft. long, 14½ in. wide?

```
16×14½=232,   12)232
              Ans. 19⅓ ft.
```

See notes, page 37.

To measure Scantlings, Joists, Plank, Sills, etc.

RULE.—*Multiply the width, the thickness, and the length together* (the width and thickness in inches, and the length in ft.), *and divide the product by* 12—*the result will be square feet.*

EXAMPLES.—How many square ft. in a Scantling 2 by 4, 16 ft. long?	$2\times4\times16=128$, 12) 128 Ans. $10\frac{2}{3}$ ft.
In a Scantling 4 by 4, 18 ft. long?	$4\times4\times18=288$, 12) 288 Ans. 24 ft.
In a Joist 2 by 8, 16 ft. long?	$2\times8\times16=256$, 12) 256 Ans. $21\frac{1}{3}$ ft.
In a Plank $2\frac{1}{2}$ by 14, 18 ft. long?	$2\frac{1}{2}\times14\times18=630$, 12) 630 Ans. $52\frac{1}{2}$ ft.
In a Sill 8 by 8, 14 ft. long?	$8\times8\times14=896$, 12) 896 Ans. $74\frac{2}{3}$ ft.

Land Measure.

To find the number of Acres in a body of Land.

RULE.—*Multiply the length by the width* (in rods), *and divide the product by* 160 (carrying the division to 2 decimal places if there is a remainder); *the result will be the answer in acres and hundredths.*

EXAMPLES.—How many Acres in a field 90 rods long and 80 rods wide?

```
         90
         80
16|0) 720|0 (45  Ans.
         80
```

How many acres in a pasture 58 rods long, and $37\frac{1}{2}$ wide?

We carry the division to two decimal places, the answer is then 13 acres and 59 hundredths of an acre.

```
5 8 × 3 7 ½ = 2 1 7 5 square rods.

1 6|0) 2 1 7|5 (1 3.5 9 + Ans.
         5 7
          9 5
          1 5 0
             6 Rem.
```

When the opposite sides of a piece of land are of unequal length, *add them together and take one-half for the mean length or width*, as will be illustrated by the following example.

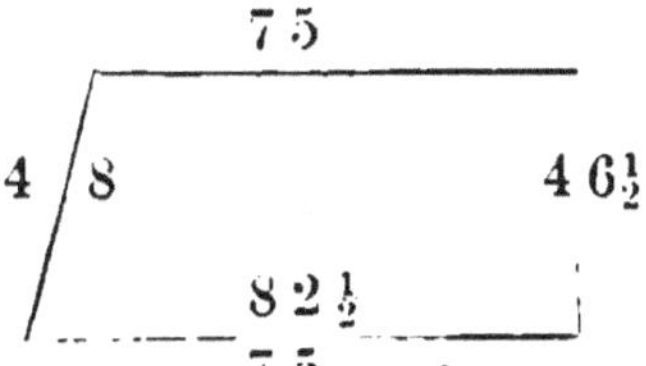

```
          75
    ______________________
4  /8                     46½
  /______82½______________
          75
      2)157.5
         78.75

         46½
         48
      2)94.5
         47.25

      78.75  mean length.
     52.74     "  width reversed.
     3150
      551
       20
160) 3721 (23.25+ Ans.
      52
       41
        90
        10 Rem.
```

This is not strictly according to geometrical principles, but is sufficiently accurate for practical purposes.

Floor, Wall, and Roof Measure.

To find the number of *Square Yards* in a Floor or Wall.

RULE.—*Multiply the length by the width or height* (in ft.), *and divide the product by* 9, *the result will be square yards.*

EXAMPLES.—How many square yds. in a Floor 18 ft. wide, and 20 ft. long?

```
     18
     20
  9) 360  square ft.
Ans. 40     "    yds.
```

How many yds. of carpet, $\frac{3}{4}$ of a yd. wide, will it take for a floor 16 ft. long and $15\frac{3}{4}$ ft. wide?

We reverse the length and write it with its units under the units of the width, the product will then be a whole number.

To divide 28 by $\frac{3}{4}$, we multiply it by the denominator (4), and divide the product by the numerator (3).

```
     15.75  width.
       61   length reversed.
     158
      94
  9) 252    square ft.
      28      "    yds.
       4
  3) 112
Ans. 37⅓  yds.
```

What will the plastering of a Room 18 by 20, and 11 ft. high, cost at 15 cts. per sq. yd? The length of the walls is 76 ft.

$76 \times 11 =$ 836 sq. ft. in 4 walls.
$18 \times 20 =$ 360 " " ceiling.
9) 1196
133 sq. yds. nearly.
15
Ans. \$19.95 cost.

To find the number of Bricks required in a building.

RULE.—*Multiply the number of cubic feet by* $22\frac{1}{2}$.

The number of cu. ft. is found by multiplying the length, height, and thickness (in. ft.) together.

Bricks are usually made 8 in. long, 4 inches wide, and 2 in. thick; hence, it requires 27 bricks to make a cu. ft. without mortar, but it is generally assumed that the mortar fills $\frac{1}{6}$ of the space.

EXAMPLES.—How many bricks are required to pave a walk 78 ft. long and 6 ft. wide, reckoning $4\frac{1}{2}$ bricks to the sq. ft.; and what will they cost at \$7.50 per thousand?

$78 \times 6 = 468$ sq. ft.
$4\frac{1}{2}$
Ans. 2106 bricks.
$7\frac{1}{2}$
Ans. \$15.795 cost.

How many bricks are required for a House whose walls are 156 ft. long, 20 ft. high, and $1\frac{1}{3}$ ft. (16 in.) thick; deducting 640 cu. ft. for doors and windows?

$156 \times 20 \times 1\frac{1}{3} = 4160$ cu. ft.
640
3528
$22\frac{1}{2}$
Ans. 79200 bricks.

How many bricks will it take to wall up a Cellar, 17 by 18, $6\frac{1}{2}$ ft. high, with an 8 inch ($\frac{2}{3}$ ft.) wall; and how many to pave the floor, reckoning $4\frac{1}{2}$ brick to the square foot?

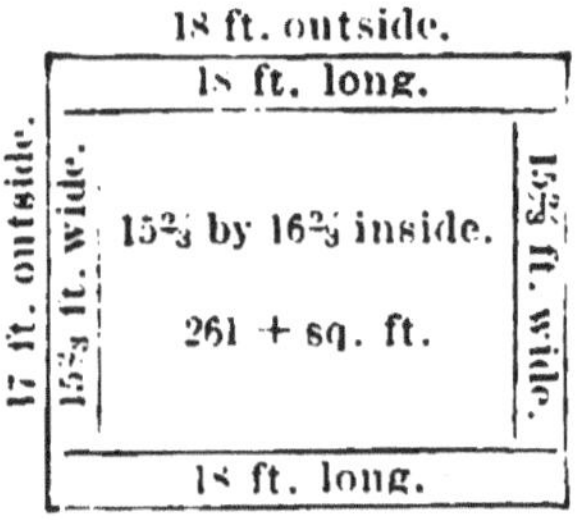

$67\frac{1}{3} \times 6\frac{1}{2} \times \frac{2}{3} =$ 292 cu. ft., nearly.
$22\frac{1}{2}$
6570 bricks in walls.

$16\frac{2}{3} \times 15\frac{2}{3} \times 4\frac{1}{2} = 1175$ bricks in floor.
Ans. 7745 bricks in all.

The whole length of the wall is $67\frac{1}{3}$ ft., viz., twice 18 ft. and twice $15\frac{2}{3}$ ft.

To find the number of Shingles required in a roof.

RULE.—*Multiply the number of square feet in the roof by 8, if the shingles are exposed 4½ inches, or by 7⅕ if exposed 5 inches.*

To find the number of square feet, *multiply the length of the roof by twice the length of the rafters.*

To find the length of the rafters, at *one-fourth* pitch, *multiply the width of the building by* .56 (hundredths); at *one-third* pitch, by .6 (tenths); at *two-fifths* pitch, by .64 (hundredths); at *one-half* pitch, by .71 (hundredths). This gives the length of the rafters from the apex to the end of the wall, and whatever they are to project must be taken into consideration.

NOTE.—By ¼, or ⅓ pitch is meant that the apex or comb of the roof is to be ¼ or ⅓ the width of the building *higher* than the walls or base of the rafters.

EXAMPLES.—How many shingles will it take to cover a shed, the roof of which is 21 ft. long and 15 ft. wide, reckoning 7⅕ shingles to the sq. ft.; and what will they cost at $4.25 per thousand?

21 × 15 = 315 sq. ft.
7⅕
Ans. 2268 shingles.
4¼
Ans. $9.639 cost.

How many shingles will it take to cover a roof 34 ft. long, and 27½ ft. from eave to eave. The shingles to be exposed 5 inches?

34 × 27½ = 935 sq. ft.
7⅕
Ans. 6732

How many shingles are required to cover a building 42 feet long, and 30 feet wide; the roof to have ⅓ pitch, and to project 1 foot on each end, and 1 foot on each side for the eaves—the shingles to be exposed 4½ inches to the weather?

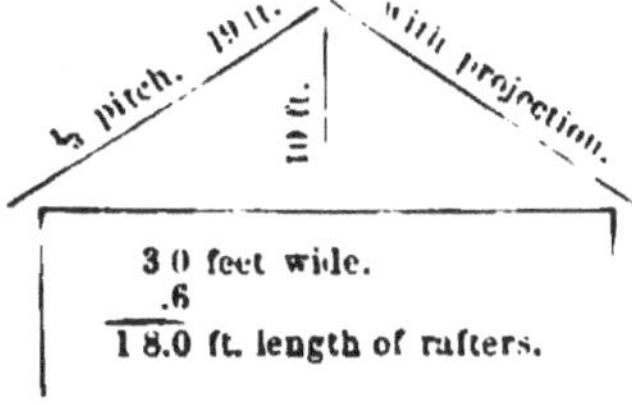

2 times 19 — 38
42 and 2 — 44
1672 sq. ft.
8
Ans. 13376

ACCOUNTS.

Every Farmer and Mechanic, whether he does much or little business, should keep a record of his transactions in a clear and systematic manner. For the benefit of those who have not had the opportunity of acquiring a primary knowledge of the principles of book-keeping, we here present a simple form of keeping accounts which is easily comprehended, and well adapted to record the business transactions of farmers, mechanics, and laborers.

1873.		ALBERT DAVIS.	Dr.		Cr.	
Jan.	10	To 7 bu. Wheat.........@ 1.25	8	75		
"	17	By shoeing span of Horses......			2	50
Feb.	4	To 14 bu. Oats............@ .45	6	30		
"	4	" 5 lbs. Butter...........@ .25	1	25		
Mar.	8	By new Harrow....................			18	00
"	8	" sharpening 2 Plows..........				40
"	13	" new Double-tree..............			2	25
"	27	To Cow and Calf...	48	00		
Apr.	9	" half ton of Hay..............	6	25		
"	9	By Cash			25	00
May	6	" repairing Corn-planter......			4	75
"	24	To one Sow with pigs............	17	50		
July	4	By Cash, to balance account...			35	15
			$88	05	$88	05

1873.		HENRY EDWARDS.	Dr.		Cr.	
Mar.	21	By 3 days' Labor.........@ 1.25			3	75
"	21	To 2 Shoats .... " 3.00	6	00		
"	23	" 18 bu. Corn......... " .45	8	10		
May	1	By 1 month's Labor.............			25	00
"	1	To Cash	10	00		
June	19	By 8 days' Mowing.......@ 1.50			12	00
"	26	To 50 lbs. Flour....................	2	75		
July	10	" 27 lbs. Meat..........@ .10	2	70		
"	29	By 9 days' Harvesting... " 2.00			18	00
Aug.	12	" 6 days' Labor.......... " 1.50			9	00
"	12	To Cash..............................	20	00		
Sept.	1	" " to balance account....	18	20		
			$67	75	$67	75

APPENDIX.

Simultaneous, or Cross Multiplication.

By this method of multiplication the product of any two numbers may be obtained without making any figures except the product itself. It is, however, a somewhat difficult process to explain it thoroughly with the pen alone. In practice, the work is really much simpler and less tedious than it appears on paper, for then we name results only and thereby obviate a considerable portion of the labor.

We present this method for the benefit of intelligent students, knowing it to be well adapted to drill and expand the mental powers. It may also be applied with advantage to practical calculations by a good accountant, and besides it is a great satisfaction to any one who thoroughly understands its principles.

RULE.—*First multiply the units together, then multiply the figures which produce tens, and adding the products mentally, set down the result and carry as usual.*

Next multiply the figures which produce hundreds, and add the products as before.

In like manner perform the multiplications which produce thousands, etc., adding the products of each order as you proceed, and thus continue the operation till all the figures are multiplied.

EXAMPLES.—Multiply 78 by 53.

```
        7 8
        5 3
Ans. 4 1 3 4
```

First we multiply the unit's figures 3 and 8 together, making 24; we set down the 4 and carry 2 (tens). Next we multiply the ten's fig. 7 by the unit's fig. 3, and the unit's fig. 8 by the ten's fig. 5, and add the two products together mentally, making 63 with the 2 (tens); we set down the 3 and carry the 6. We then multiply the ten's fig. 7 by the ten's fig 5, which with the 6 (tens) makes 41.

Multiply 354 by 62.

	354
	62
Ans.	21948

First we multiply the 4 units by the 2 units. Second, the 5 tens by the 2 units, and the 4 units by the 6 tens, making 34. Third, the 3 hundreds by the 2 units, and 5 tens by the 6 tens, making 39 with the 3 to carry. Fourth, the 3 hundreds by the 6 tens, making 21, including the 3 (tens).

Multiply 627 by 453.

	627
	453
Ans.	284031

First we multiply the 7 by the 3. Second, the 2 by the 3 and the 7 by the 5, making 43, including the tens. Third, the 6 by the 3, the 2 by the 5 and the 7 by the 4, making 60. Fourth, the 6 by the 5 and the 2 by the 4, making 44. Fifth, the 6 by the 4, making 28.

Multiply 7325 by 614.

	7325
	614
Ans.	4497550

First say, 4 times 5 are 20. Second, 2 to carry to 4 times 2 and 1 time 5, make 15. Third, 1 to carry to 4 times 3, 1 time 2 and 6 times 5, make 45. Fourth, 4 to carry to 4 times 7, 1 time 3 and 6 times 2, make 47. Fifth, 4 to carry to 1 time 7 and 6 times 3, make 29. Sixth, 2 to carry to 6 times 7 make 44.

Multiply 4587 by 3126.

	4587
	3126
Ans.	14338962

First say, $6 \times 7 = 42$. Second, 4 (tens), 6×8 and $2 \times 7 = 66$. Third, 6 (tens), 6×5, 2×8 and $1 \times 7 = 59$. Fourth, 5 (tens), 6×4, 2×5, 1×8 and $3 \times 7 = 68$. Fifth, 6 (tens) to 2×4, 1×5 and $3 \times 8 = 43$. Sixth, 4 (tens), 1×4 and $3 \times 5 = 23$. Seventh, 2 (tens), $3 \times 4 = 14$.

Multiply 93612 by 84075.

	93612
	84075
Ans.	7870428900

First, $5 \times 2 = 10$. Second, 1 (ten), 5×1 and $7 \times 2 = 20$. Third, 2 (tens), 5×6, 7×1 and $0 \times 2 = 39$. Fourth, 3 (tens), 5×3, 7×6, 0×1 and $4 \times 2 = 68$. Fifth, 6 (tens), 5×9, 7×3, 0×6, 4×1 and $8 \times 2 = 92$. Sixth, 9 (tens), 7×9, 0×3, 4×6 and $8 \times 1 = 104$. Seventh, 10 (tens), 0×9, 4×3 and $8 \times 6 = 70$. Eighth, 7 (tens), 4×9 and $8 \times 3 = 67$. Ninth, 6 (tens) and $8 \times 9 = 78$.

Peculiar and Useful Contractions in Multiplication.

To Multiply any number of two figures by 11. *Write the sum of the two figures between them.*

Multiply 34 by 11. Say 3 and 4 are 7, and write it between the 3 and 4. Ans. 374.

Multiply 97 by 11. Say 9 and 7 are 16, write the 6 in the middle, and add the 1 to the 9. Ans. 1067.

To find the product of any two numbers, whose *right* hand figures make 10, and whose *left* hand figures are *alike.*

Multiply the units together and set down their product, then add 1 to the upper tens and multiply it by the lower, and set their product before the product of the units.

Multiply 75 by 75.

Say 5 times 5 are 25 and set it down, then increase the upper 7 by 1, and say 7 times 8 are 56, which set before the 25.

```
       7 5
       7 5
Ans. 5 6 2 5
```

Multiply 117 by 113.

Say 3 times 7 are 21; add 1 to 11 and say 11 times 12 are 132.

```
      1 1 7
      1 1 3
Ans. 1 3 2 2 1
```

Multiply 89 by 81.

Say once 9 is 9, set it down and prefix a 0, then say 8 times 9 are 72.

```
       8 9
       8 1
Ans. 7 2 0 9
```

When the *left* hand figures make 10, and the *right* hand figures are *alike.*

Set down the product of the units, and to the left of it the product of the tens INCREASED *by one of the units figures.*

Multiply 58 by 58.

Say 8 times 8 are 64 and set it down, then say 5 times 5 are 25 and 8 (one of the units), make 33.

```
       5 8
       5 8
Ans. 3 3 6 4
```

Multiply 62 by 42.

Say 2 times 2 are 4, set it down and prefix a 0, then say 4 times 6 are 24 and 2 make 26.

```
       6 2
       4 2
Ans. 2 6 0 4
```

To square any number of 9s instantaneously, and without making any figures except the product itself.

Begin on the left and write as many 9s, less one, as there are 9s in the given number, an 8, as many 0s as 9s, and a 1.

What is the square of 999? Ans. 998001.
Set down two 9s, an 8, two 0s, and a 1.

Find the square of 99999. Ans. 9999800001.
Here are five 9s, write four 9s, an 8, four 0s, and a 1.

Contracted Division—A New Method.

By this method of division which is scientific and practical, the quotient is obtained by an easy process with very few figures, and far less labor than would at first be inferred from the rule. It possesses the peculiar characteristic, that the larger the divisor, the less figures and labor is required in the operation. The diligent student will never regret the time and labor he bestows, in trying to learn and comprehend the principles of this useful and amusing method.

RULE.—*Assume as many figures of the dividend as will contain the integral part of the divisor, count the remaining figures in the integral part of the dividend, which,* INCREASED *by* 1, *will be the number of figures in the integral part of the quotient. If the division is to be carried to decimals, increase this number by as many as there will be decimal places in the quotient.*

Take as many figures of the divisor as there will be figures in the quotient, annexing ciphers if there are not as many. Take as many figures of the dividend as will contain this divisor, and if there are not enough, supply the deficiency by annexing ciphers.

Obtain the first quotient figure in the usual manner, multiply the divisor by this figure, carrying the tens, however, from the nearest rejected figure in the divisor, and write only the remainders in the same manner as in "Short Method of Division."

Reject the right hand figure of the preceding divisor and use the last remainder for the next partial dividend, and thus proceed until the divisor is reduced to a single figure, then point off the required number of decimals.

EXAMPLES.—Divide 4972356 by 21345, carrying the division to units.

Assuming as many figures as will contain the
```
2,1,3|4 5) 4 9 7|2 3 5 6 (2 3 3 Ans., nearly.
                 7 0
                   6
```
whole divisor, there are *two* figures remaining in the dividend, by this we know that there will be *three* figures in the quotient.

We now take the three left hand figures of the divisor, and as many of the dividend as will contain them, and proceed thus, 213 in 497 is contained 2 times; setting the 2 in the quotient we say, 2 times 3 are 6 and 1 (ten) from the nearest rejected fig. 4, makes 7, which would fall under the 7 in the dividend, but writing the remainders only, we set a 0 in its place. We then say, 2 times 1 are 2 and 7 (written in the rem.) make 9; 2 times 2 are 4, (no rem.)

We now mark off the 3 in the divisor and say, 21 in 70, 3 times, 3 times 1 are 3, and 1 (ten) from the rejected fig. 3, makes 4 and 6 (written in the rem.), make 10; 3 times 2 are 6 and 1 (ten), makes 7, (no rem.) We next mark off the 1 in the divisor and say, 2 in 6, 3 times, 3 times 2 are 6, (no rem.), which finishes the operation.

Divide 523824 by 748, carrying the division to 1 decimal place.

```
7,4,8,0) 5 2 3 8 2 4 (7 0 0.3 Ans.
                 2 2
```

Here there are 2 figures left in the dividend after assuming enough to contain the divisor (748), hence, there will be 3 figures in the integral part, and with the 1 decimal —4 places in the quotient. There being only 3 figures in the divisor, we annex a 0 to it; take as many figures of the dividend as will contain it now, and proceed thus; 7480 in 52382, 7 times and 22 over, we then mark off the 0 in the divisor and say, 748 in 22, 0 time, set a 0 in the quotient and mark off the 8, and proceed, 74 in 22, 0 time, set another 0 in the quotient, mark off the 4 and say, 7 in 22, 3 times, 3 times 7 are 21 and 1 (ten from the rejected figure 4, make 22, (no rem.)

Divide 8186352.9375 by 3967.3125, carrying the division to 2 decimal places.

Comparing the integral part of the divisor with the integral part of the dividend, shows that there will be 4 figures in the integral part of the quotient, and with the 2 decimals—6 places in all. We then take the first 6 figures of the divisor, and as many of the dividend as will contain them and proceed as before.

```
3,9,6,7,.3,1|2 5) 8 1 8 6 3 5|2.9 3 7 5 (2 0 6 3.4 5 Ans.
                    2 5 1 7 3
                      1 3 6 9
                        1 7 9
                          2 0*
```

See notes, page 37.

 Table showing value of articles sold by the piece, pound, yd. or doz.; as Butter, Eggs, &c. see Explanations, p. 6.

1	2	3	4	5	6	7	8	9	10	11	12	13	14	15	16	17	18	19		20	21	22	23	24	25	26	27	28	29	30	31	32	35
12	24	36	48	60	72	84	96	108	120	132	144	156	168	180	192	204	216	228	12	240	252	264	276	288	300	312	324	336	348	360	372	384	420
13	26	39	52	65	78	91	104	117	130	143	156	169	182	195	208	221	234	247	13	260	273	286	299	312	325	338	351	364	377	390	403	416	455
14	28	42	56	70	84	98	112	126	140	154	168	182	196	210	224	238	252	266	14	280	294	308	322	336	350	364	378	392	406	420	434	448	490
15	30	45	60	75	90	105	120	135	150	165	180	195	210	225	240	255	270	285	15	300	315	330	345	360	375	390	405	420	435	450	465	480	525
16	32	48	64	80	96	112	128	144	160	176	192	208	224	240	256	272	288	304	16	320	336	352	368	384	400	416	432	448	464	480	496	512	560
17	34	51	68	85	102	119	136	153	170	187	204	221	238	255	272	289	306	323	17	340	357	374	391	408	425	442	459	476	493	510	527	544	595
18	36	54	72	90	108	126	144	162	180	198	216	234	252	270	288	306	324	342	18	360	378	396	414	432	450	468	486	504	522	540	558	576	630
19	38	57	76	95	114	133	152	171	190	209	228	247	266	285	304	323	342	361	19	380	399	418	437	456	475	494	513	532	551	570	589	608	665
20	40	60	80	100	120	140	160	180	200	220	240	260	280	300	320	340	360	380	20	400	420	440	460	480	500	520	540	560	580	600	620	640	700
21	42	63	84	105	126	147	168	189	210	231	252	273	294	315	336	357	378	399	21	420	441	462	483	504	525	546	567	588	609	630	651	672	735
22	44	66	88	110	132	154	176	198	220	242	264	286	308	330	352	374	396	418	22	440	462	484	506	528	550	572	594	616	638	660	682	704	770
23	46	69	92	115	138	161	184	207	230	253	276	299	322	345	368	391	414	437	23	460	483	506	529	552	575	598	621	644	667	690	713	736	805
24	48	72	96	120	144	168	192	216	240	264	288	312	336	360	384	408	432	456	24	480	504	528	552	576	600	624	648	672	696	720	744	768	840
25	50	75	100	125	150	175	200	225	250	275	300	325	350	375	400	425	450	475	25	500	525	550	575	600	625	650	675	700	725	750	775	800	875
26	52	78	104	130	156	182	208	234	260	286	312	338	364	390	416	442	468	494	26	520	546	572	598	624	650	676	702	728	754	780	806	832	910
27	54	81	108	135	162	189	216	243	270	297	324	351	378	405	432	459	486	513	27	540	567	594	621	648	675	702	729	756	783	810	837	864	945
28	56	84	112	140	168	196	224	252	280	308	336	364	392	420	448	476	504	532	28	560	588	616	644	672	700	728	756	784	812	840	868	896	980
29	58	87	116	145	174	203	232	261	290	319	348	377	406	435	464	493	522	551	29	580	609	638	667	696	725	754	783	812	841	870	899	928	1015
30	60	90	120	150	180	210	240	270	300	330	360	390	420	450	480	510	540	570	30	600	630	660	690	720	750	780	810	840	870	900	930	960	1050
31	62	93	124	155	186	217	248	279	310	341	372	403	434	465	496	527	558	589	31	620	651	682	713	744	775	806	837	868	899	930	961	992	1085
32	64	96	128	160	192	224	256	288	320	352	384	416	448	480	512	544	576	608	32	640	672	704	736	768	800	832	864	896	928	960	992	1024	1120
33	66	99	132	165	198	231	264	297	330	363	396	429	462	495	528	561	594	627	33	660	693	726	759	792	825	858	891	924	957	990	1023	1056	1155
34	68	102	136	170	204	238	272	306	340	374	408	442	476	510	544	578	612	646	34	680	714	748	782	816	850	884	918	952	986	1020	1054	1088	1190
35	70	105	140	175	210	245	280	315	350	385	420	455	490	525	560	595	630	665	35	700	735	770	805	840	875	910	945	980	1015	1050	1085	1120	1225
36	72	108	144	180	216	252	288	324	360	396	432	468	504	540	576	612	648	684	36	720	756	792	828	864	900	936	972	1008	1044	1080	1116	1152	1260
37	74	111	148	185	222	259	296	333	370	407	444	481	518	555	592	629	666	703	37	740	777	814	851	888	925	962	999	1036	1073	1110	1147	1184	1295
38	76	114	152	190	228	266	304	342	380	418	456	494	532	570	608	646	684	722	38	760	798	836	874	912	950	988	1026	1064	1102	1140	1178	1216	1330
39	78	117	156	195	234	273	312	351	390	429	468	507	546	585	624	663	702	741	39	780	819	858	897	936	975	1014	1053	1092	1131	1170	1209	1248	1365
40	80	120	160	200	240	280	320	360	400	440	480	520	560	600	640	680	720	760	40	800	840	880	920	960	1000	1040	1080	1120	1160	1200	1240	1280	1400
41	82	123	164	205	246	287	328	369	410	451	492	533	574	615	656	697	738	779	41	820	861	902	943	984	1025	1066	1107	1148	1189	1230	1271	1312	1435
42	84	126	168	210	252	294	336	378	420	462	504	546	588	630	672	714	756	798	42	840	882	924	966	1008	1050	1092	1134	1176	1218	1260	1302	1344	1470
43	86	129	172	215	258	301	344	387	430	473	516	559	602	645	688	731	774	817	43	860	903	946	989	1032	1075	1118	1161	1204	1247	1290	1333	1376	1505
44	88	132	176	220	264	308	352	396	440	484	528	572	616	660	704	748	792	836	44	880	924	968	1012	1056	1100	1144	1188	1232	1276	1320	1364	1408	1540
45	90	135	180	225	270	315	360	405	450	495	540	585	630	675	720	765	810	855	45	900	945	990	1035	1080	1125	1170	1215	1260	1305	1350	1395	1440	1575

www.ingramcontent.com/pod-product-compliance
Lightning Source LLC
Chambersburg PA
CBHW021958190326
41519CB00009B/1309

* 9 7 8 3 3 3 7 3 8 9 3 9 0 *